# ENGLISH FOR MARITIME STUDIES

**SECOND EDITION**

**_Other_ English For Specific Purposes _Titles_**

ADAMSON, D.
*International Hotel English**

ADAMSON, D.
*Starting English for Business**

BEECH, J.
*Thank You for Flying With Us**

BRIEGER, N. et al.
*Business Contacts**

BRIEGER, N and COMFORT, J.
*Business Management English (Series)**

BRIEGER, N and COMFORT, J.
*Developing Business Contacts**

BRIEGER, N and COMFORT, J.
*Early Business Contacts**

BRIEGER, N and COMFORT, J.
*Social Contacts**

BRIEGER, N and COMFORT, J.
*Technical Contacts**

BRIEGER, N and CORNISH, A.
*Secretarial Contacts**

HICK, S.
*English for Information Systems**

KEANE, L.
*International Restaurant English**

MCGOVERN, J. and MCGOVERN, J.
*Bank on Your English**

PALSTRA, R.
*Telephone English**

PALSTRA, R.
*Telex English*

POTE, M. et al.
*A Case for Business English**

ROBERTSON, F.
*Airspeak**

*Includes audio cassettes

# ENGLISH FOR MARITIME STUDIES

**SECOND EDITION**

T. N. Blakey

**Pearson Education Limited**
Edinburgh Gate, Harlow
Essex CM20 2JE
and Associated Companies throughout the World

www.longman-elt.com



ISBN: 0.13.281379.3

First published in 1983 by Pergamon Press
This edition published by Pearson Education Limited 1999

Third impression 2001



Printed in China
NPCC/03

# CONTENTS

## UNIT III Ship Construction



## UNIT IV Manning



## UNIT V Seamanship

## UNIT VI Cargo Work



## UNIT VII Navigation



## UNIT VIII Main Engines

## UNIT IX Auxiliary Machinery



## UNIT X Maintenance



## UNIT XI Safety Aboard



## UNIT XII Communications at Sea

# ACKNOWLEDGEMENTS


The author and publisher would like to thank the following for their kind permission to reproduce extracts and photographs:

**United Arab Shipping Company, S.A.G.:** general cargo carrier; container ship.

**Ben Line:** dry bulk carrier; liquid bulk carrier; containers being loaded; general cargo carrier; loading a bus; unloading coal.

**P&O Ferries:** ferry.

**Henry Robb Ltd., Shipbuilders, Leith:** grab dredger; stockyard; preparation shop; prefabrication shop; building berth.

**B.P. Oil Ltd.:** tanker.

**Forth Ports Authority:** dry bulk carrier unloading.

**The Motor Ship Journal (IPC Business Press Ltd.):** fold-out diagram of general cargo carrier.

**Marconi International Marine Company Ltd.:** radio room, Marconigram.

**Her Majesty's Stationery Office:** Collision Regulations and Distress Signals and Code of Safe Working Practices for the Safety of Merchant Seamen.

**John Grant-Robertson (GR Language Services):** navigating instruments; gyro compass; sextant; magnetic compass; stockyard; preparation shop; prefabrication shop; building berth; dry bulk carrier unloading.

**The Royal Navy:** nuclear submarine; modern destroyer.

Grateful thanks to Captain Peter Young, Eric H. Glendinning, Rosalind Grant-Robertson and the students of the Edinburgh Language Foundation and Leith Nautical College for their help and assistance in making this work possible. Also to Thomas E. Scambler, Keith Mitchell, Captain B. Farrell and Captain F. Weeks for their help in the preparation of this second edition.

# INTRODUCTION

## Who the course is for

*English for Maritime Studies* is an English language course designed to help non-native English-speaking cadets and qualified personnel who are studying for professional qualifications at nautical colleges and naval establishments throughout the world where the medium of English is used.

The course consists of 14 units:

| | | | |
|---|---|---|---|
| I | Shipping | VIII | Main engines |
| II | Ship types | IX | Auxiliary machinery |
| III | Ship construction | X | Maintenance |
| IV | Manning | XI | Safety aboard |
| V | Seamanship | XII | Communications at sea |
| VI | Cargo work | XIII | Radio communications |
| VII | Navigation | XIV | Shipboard electronics |

Each unit covers language work relevant to the topic of that unit. Answers to the exercises and tapescripts of lectures are given in the Appendices.

The course is therefore suitable for students studying to become, or studying for further qualifications as Deck, Engineer, or Radio Officers in warships and merchant vessels. The first four units and units X and XI cover topics relevant to all departments. If time is limited, it is possible to work through the six units mentioned above plus those relevant to the student's particular department. However, this is a language course and there are useful structures and vocabulary to be learnt by working through all the units. At the same time some knowledge of aspects of work in other departments would be gained. This would be particularly beneficial for students at the pre-selection stage.

The course is designed to be used either in the classroom with a teacher or by students individually on a self-study basis. Instructions and notes to guide the teacher and student are given below and throughout the course.

The course assumes an intermediate knowledge of English and is therefore not suitable for complete beginners. If the student is following an English language course at a language school before going on to nautical college, *English for Maritime Studies* can be used in conjunction with a general English language course at intermediate or post-intermediate level.

## What the course aims to do

The course aims to provide a stepping-stone to enable a student to cross from an intermediate knowledge of general English to a position where he can handle the sort of text books and instruction he will meet at college and on board ship.

In order to help the student, the course aims to develop within themes relevant to maritime studies *reading skills, writing skills, communicating skills* and *note-taking skills.* It also aims to build up the student's vocabulary of nautical and technical terms. In addition to this, it is hoped that by working through the course the student will gain a general knowledge of ships and shipping, which he may not already have.

Each unit offers about 8–10 hours work and is divided into seven sections:

## Using the course

### *Reading comprehension (A)*

This aims to develop the ability to understand the sort of texts students will meet at college. The approach to comprehension is first to activate any knowledge the student may have of a subject before he reads about it. This is done by getting the student to consider questions, pictures or diagrams before he starts to read the text. By bringing relevant ideas and words to the forefront of the student's mind (even in his own language) and by raising certain expectations, he will be in a better position to comprehend the overall structure of a passage and to understand the meaning of important words in their context. After reading the passage the student is often asked, as a check to comprehension, to complete a summary outline of the passage. This attention to the structure of passages will help him later, when he comes to write his own descriptions. Other exercises are also given. At this stage, the concern is more for comprehension of overall structure and the main ideas, rather than every single word.

### *Grammar*

This looks in more detail at words and structures used in the reading passage, or which will be used in the next section to form a more developed reading passage. This section aims at developing a knowledge of certain aspects of grammar that will be useful for both reading and writing English. In particular, the (B) part examines the different ways statements can be linked to form more complex statements.

## *Reading comprehension (B)*

The concern here is to examine more closely how words work in order to make a coherent and meaningful piece of English. The words and structures examined in the Grammar sections are used in the reading passages to show how they work in texts and exercises are given to check that the student has understood their meaning and uses. Exercises are also given to help the student realize how sentences and paragraphs are linked on a grammatical and semantic basis to give a passage coherence and unity. This insight will also help him with his writing. In order to show that passages raise questions as well as answer them, the student is sometimes asked to use his imagination to try and answer questions not answered in the text. It is hoped that these questions will stimulate discussion and a desire for further reading.

## *Applied terminology*

This aims to build up some of the vocabulary the student will need in his particular area of study. The number of nautical and technical terms is enormous and only some have been given here. These have been chosen because of their relevance to the topic of the unit in which they appear. In the same way, part (B) deals with the language of measurement relevant to that unit.

## *Guided writing (A)*

This is for the student to use the words and structures he has learnt in a guided composition. He is guided in how to put statements together to form more complex sentences and taught how to give his composition an overall structure. It is hoped that by the end of the course, the student will be in a better position to write essays for college work and examinations.

## *Note-taking practice*

This is aimed at training the student to take meaningful notes during lectures and lessons. At each stage, new techniques are introduced and thoroughly practised, before the student is exposed to a short lecture. The practices and lectures are recorded on cassette and the tapescripts can be found in Appendix II at the back of the book. The listening and note-taking tasks are explained in each section. The symbol 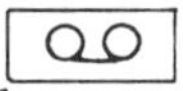 marks where the cassette should be used. As comprehension is an essential part of note-taking, it is hoped that the skills practised in the reading passages will be further developed here. The note-taking lectures also give further opportunity for increasing the student's knowledge of his subject.

### *Guided writing (B)*

This gives the student the opportunity of putting the notes he has taken in the previous section to practical and meaningful use. It will also give him the opportunity of seeing how adequate his notes are. By turning to the tapescripts, he will be able to compare his composition with the original.

## How the course might be supplemented

The course itself can be made more instructive by using pictures, photographs, slides, diagrams and realia to help with vocabulary and supplementary language work. These can be chosen for their relevance to the topic of each unit. If possible, visits to ports and shipyards can be geared to units in order to underline the practicality of the course and increase general knowledge of ships and shipping.

# UNIT I

# SHIPPING

## READING COMPREHENSION

### (A) How merchant ships operate (version 1)

The introductory paragraph to a reading passage will often tell you what the rest of the passage is going to be about. *Read the first paragraph only of the passage below and then write down a short statement saying what the passage is going to be about. Begin your statement:*

The passage is about . . .

*Now read the passage through rapidly to see if your prediction is right. Do not stop at words you do not understand.*

[1]Merchant ships are designed to carry cargo. [2]Some are also designed to carry passengers. [3]Nowadays most merchant ships are built to carry cargo. [4]A few still carry passengers. [5]Merchant vessels can operate in the following three basic ways.

[6]They can operate as liners. [7]Liners are employed on regular routes on a fixed timetable. [8]A list of their arrival and departure dates is published in advance. [9]They sail whether they are full or not.

[10]Merchant vessels can also operate as tramps. [11]Tramps do not sail on regular routes. [12]They do not keep to a fixed timetable. [13]They are employed in all parts of the world where there is cargo for them to carry.

[14]A large number of merchant ships operate as specialized vessels. [15]These are designed to carry a particular type of cargo.

Is your statement similar to the heading of the summary outline below?

*Complete the outline to form a summary of the passage.*

The three basic ways merchant vessels can operate:

(a) ............................................................

(b) ..............................................................

(c) ..............................................................

This outline also shows the structure of the passage.

Now read the passage more carefully. Underline the words you do not understand. Before you use your dictionary or ask your teacher, remember that the meaning of a word can often be got (a) from the context (e.g. 'vessels', sentence 5, means 'ships'), (b) by studying the word to see if it is derived from a noun or verb etc. that you already know (e.g. 'arrival', sentence 8, from 'to arrive').

*Use the information in the reading passage to complete these sentences.*

(a) Merchant ships are designed to carry..............................................

(b) Liners are employed..............................................................

(c) Tramps are not employed.........................................................

(d) Specialized vessels are designed...............................................

# GRAMMAR

## (A) Quantifiers

Quantifiers, or amount words, indicate the approximate quantity or amount of the noun they qualify or represent.

*Study the table below which gives the most commonly used quantifiers.*

| Amount | Countables | Uncountables |
|---|---|---|
| a total amount | all | all |
| an almost total amount | most | most |
| a large amount | many<br>a lot (of)<br>a large number (of) | much<br>a lot (of)<br>a large amount (of) |
| a small amount | some<br>several<br>a few | some<br>–<br>a little |
| a very small amount | few | little |
| zero amount | no<br>none | no<br>none |

*Study the following points and examples:*

1. **All, most, many, much, some, several, (a) few, (a) little,** can be used without a noun to avoid repetition

   e.g. *All liners* carry passengers or cargo. *All* follow regular routes.
   *Some cargo* is carried in holds. *Some* is carried on deck.

   (*Note:* **Much** is normally used in only negative and interrogative sentences.
   e.g. He had not much money so he walked into town. Is there much cargo to be unloaded?)

2. When **a large number, a large amount** and **a lot** are followed by a noun, **of** must be used
   e.g. A large number *of* merchant ships carry bulk cargo.

3. **No** and **none** are used as follows:
   e.g. *No tankers* can carry passengers. *None* should discharge oil into the sea.

4. There is a **positive/negative** contrast between **a little** and **little** and between **a few** and **few**. *Study the difference in meaning in these sentence pairs:*
   There was a little time before the ship sailed, so he stayed ashore longer.
   There was little time before the ship sailed, so he went on board immediately.
   A few people were sitting on deck, because the sun was shining.
   Few people were sitting on deck, because the wind was cold.

*Exercise* 1. *Complete the table below to show which quantifiers can be used with the* ***countable*** *noun 'passengers' and the* ***uncountable*** *noun 'money'. Two have been done for you. If the noun cannot be used with a particular quantifier, you must leave a blank space.*

| Quantifier | Countable noun | Uncountable noun |
|---|---|---|
| several | ............................ | ............................ |
| no | ............................ | ............................ |
| a lot of | ............................ | money |
| few | ............................ | ............................ |
| much | ............................ | ............................ |
| a large number of | ............................ | ............................ |
| a little | ............................ | ............................ |
| all | ............................ | ............................ |
| many | passengers | ............................ |

*(continued overleaf)*

*Exercise* 1.

| Quantifier | Countable noun | Uncountable noun |
|---|---|---|
| a large amount of | .............................. | .............................. |
| a few | .............................. | .............................. |
| some | .............................. | .............................. |
| most | .............................. | .............................. |
| little | .............................. | .............................. |

*Exercise* 2. *Choose a suitable quantifier and either 'passengers' or 'money' to complete these sentences:*

..............................................travel by cargo liner.

..........................................is needed to operate a shipping fleet.

.......................................enjoy being at sea when it is rough.

..........................................are allowed down in the engine room.

...................................on board ship should be kept in a safe place.

## (B) Logical connectives (i)

Structures can be linked in a number of ways. One way is by using connecting words such as **and, or, but.**
*Study how they are used in these examples:*

1. **And** is a general purpose link
   (a) Passenger liners carry passengers.
   (b) Passenger liners carry cargo.
   (a) + (b) Passenger liners carry passengers *and* cargo.
   Here it means they carry both and suggests at the same time.

2. **Or** adds an alternative
   (a) Cargo liners are designed to carry containerized cargoes.
   (b) Cargo liners are designed to carry conventional cargoes.
   (a) + (b) Cargo liners are designed to carry containerized *or* conventional cargoes.
   If **and** were used here it would suggest 'at the same time'.

3. **But** suggests contrast or the addition of something unexpected
   - (a) Merchant ships are classified by type and trade.
   - (b) Warships are classified by type and size.
   - (a) + (b) Merchant ships are classified by type and trade, *but* warships are classified by type and size.

*Exercise* 1. *Join the following groups of sentences using* **and, or, but,** *as appropriate:*

(a) A cadet can train as a Deck Officer.
A cadet can train as an Engineer Officer.

(b) The ship was old.
The ship was in good condition.

(c) Passenger liners carry passengers.
Passenger liners carry some cargo.

(d) Tankers usually operate as specialized vessels.
Tankers sometimes operate as tramps.

(e) Merchant ships are designed to carry liquid cargo.
Merchant ships are designed to carry dry cargo.

(f) Warships are designed for speed.
Warships are designed for manoeuvrability.

Sometimes the linking of two structures can be made more complete by putting a word before the connecting word. This tells us to expect an *addition* or an *alternative* a little later. *Study these samples.*

1. Additions: **both . . . and; not only . . . but also**
   Some ferries carry *both* passengers *and* cars.
   Some ferries carry *not only* passengers *but also* cars.

2. Alternatives: **either . . . or; neither . . . nor** (negative)
   Merchant ships are designed to carry *either* liquid *or* dry cargo.
   The sailor had *neither* money *nor* his passport.

*Exercise* 2. *Rewrite these sentences using the above pairs of connecting words as appropriate in as many ways as possible, but keeping the original meaning:*

(a) He failed the practical exam and the written exam.
(b) A cadet can train as a Deck or an Engineer Officer.
(c) The crew saved the ship and the cargo.
(d) They had no food and no water for two days.
(e) Some ferries carry passengers and vehicles.
(f) Merchant ships operate as tramps or liners.

# READING COMPREHENSION

## (B) How merchant ships operate (version 2)

In English, sentences and parts of sentences are linked by words and phrases which relate them to each other. For example, a noun (*ships*) in one sentence may become a pronoun (*they*) in the next; a quantifier (*some*) may be used without a noun (*merchant ships*) to avoid repetition. Also there are a number of words and phrases which refer the reader forwards to what is about to be said (*the following, below*) or backwards to what has been said (*the former, the latter, above*).

*Now read through this passage, which is a development of the reading passage in Reading Comprehension (A). As you read, circle the words and phrases which the words underlined refer to. The first has been done for you.*

[1]Merchant ships are designed to carry cargo. [2]Some are also designed to carry passengers. [3]Nowadays, most merchant ships are built to carry cargo, but a few still carry passengers. [4]Merchant vessels can operate in the following three basic ways.

[5]They can operate as liners. [6]These are employed on regular routes on a fixed timetable. [7]A list of their arrival and departure dates is published in advance and they sail whether full or not. [8]Liners can be classed as either deep-sea liners or short-sea liners. [9]The former carry mainly containerized cargo across the oceans of the world; the latter carry containerized or conventional cargo on shorter routes. [10]Ferries are also classed as liners. [11]These offer a daily or weekly service for passengers and vehicles across channels and narrow seas. [12]A few ships are still employed as passenger liners. [13]They not only carry passengers but also some cargo on routes from Europe to North America and to the Far East. [14]Nowadays the passenger trade is very small and passenger liners usually operate as cruise ships for part of the year.

[15]Merchant ships also operate as tramps. [16]These vessels do not sail on regular routes or keep to a fixed timetable, but are employed where there is cargo for them to carry. [17]Tramps can be classed as deep-sea tramps or short-sea tramps. [18]A number are classed as coasters. [19]These ply on coastal routes and up rivers to inland ports. [20]The traditional tramp cargoes are dry bulk cargoes, but some are designed to carry general cargoes.

[21]A large number of merchant ships operate as specialized vessels. [22]These are designed to carry a particular type of cargo. [23]There are several types of specialized vessel. [24]The most common are oil tankers. [25]They are owned by the major oil companies or by independent operators. [26]Two other types of liquid bulk carrier of growing importance are chemical carriers and liquefied natural gas (LNG) carriers.

The basic structure of this passage is the same as the first one, but more information has been added.

*Copy out the outline in Reading Comprehension (A), but this time add to each of the three groups the different types of ship which make up that category. After each type of vessel write in the type of cargo they carry. The first part has been started for you.*

(a) liners – deep-sea liners (containerized)
short-sea liners (containerized or conventional)

..........................................................................................

Passages not only confirm certain expectations and answer certain questions about the topic under discussion that the reader may have, they also raise questions and stimulate the imagination or a desire for further reading.

*Use your imagination to try to answer these questions:*

(a) Why do only a few merchant ships carry passengers nowadays?
(b) Why do passenger liners operate as cruise ships for part of the year?
(c) Why are tramps mainly designed to carry dry bulk cargoes?
(d) What other types of specialized vessel can you think of?

# APPLIED TERMINOLOGY

## (A) Terms relating to shapes

The shapes of objects can be referred to by using a number of methods:

1. By using the names of geometric figures

| Two-dimensional figures | Noun | Adjective | Three-dimensional figures | Noun | Adjective |
|---|---|---|---|---|---|
| 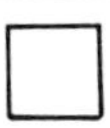 | square | square |  | cube | cubic |
| 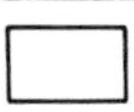 | rectangle | rectangular | 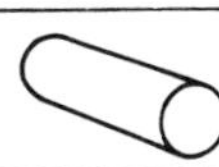 | cylinder | cylindrical |
|  | triangle | triangular |  | pyramid | pyramidal |
|  | circle | circular |  | sphere | spherical |
|  | semicircle | semicircular |  | hemisphere | hemispherical |
|  | ellipse/oval | elliptical/oval |  | cone | conical |

When used to describe shapes, the noun or adjective can be used.

e.g. It is shaped like a *circle*.
It is *circular* in shape.

*Exercise* 1. *Describe the shape of these objects using both patterns:*

(a) 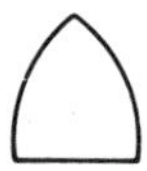
Starboard hand buoy

(b) 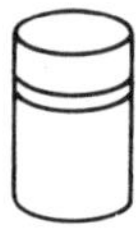
Funnel

(c) 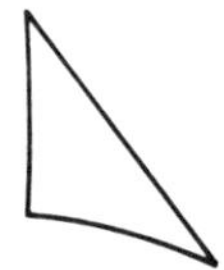
Sail

(d) 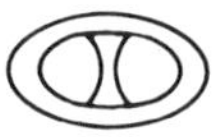
Cable link

(e) 
Special mark buoy

(f) 
Rudder

(g) 
Lifebuoy

(h) 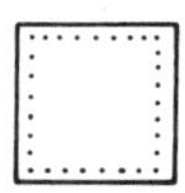
Faceplate

(i) 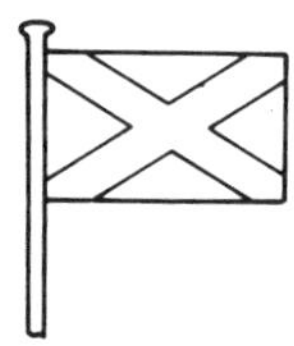
Flag

2. By using letters of the alphabet
These are used to describe shapes in the following ways:

a beam shaped like the letter I = an I-beam

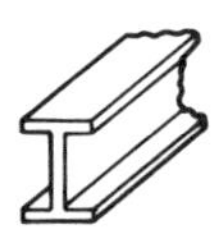

a pipe shaped like the letter S = an S-shaped pipe

3. By using objects with well-known shapes

| | | |
|---|---|---|
| e.g. an eye | a mushroom | a needle |
| a heart | a kidney | a finger |

Can you suggest any others?
These are used as follows:
a ventilator shaped like a mushroom = a mushroom ventilator
a valve shaped like a mushroom = a mushroom-shaped valve

*Exercise* 2. *Complete these names using a letter of the alphabet or one of the objects above:*

(a)  (beam)

(b)  (splice)

(c)  (bracket)

(d)  (shackle)

(e) 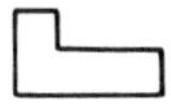 (cabin) (use '-shaped')

(f)  (ventilator)

(g)  (valve)

(h) 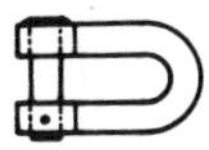 (shackle)

(i)  (plates)

(j)  (cut) (use '-shaped')

(k)  (tray)

(l)  (tube) (use '-shaped')

# (B) Measurement

## *(i) General dimensions*

We can describe the length of an object using four different patterns:

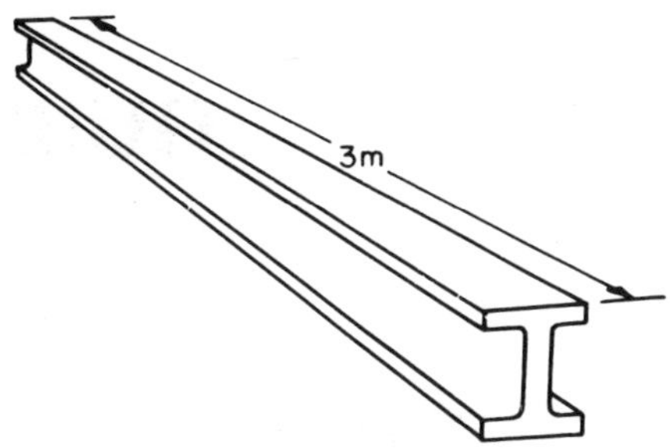


1. The beam is three metres in length.
2. The beam has a length of three metres.
3. The length of the beam is three metres.

or by using the adjective

4. The beam is three metres long.

**Width/breadth, height, depth** and **thickness** are described using the same patterns. **Radius, diameter, circumference** are described using patterns 1, 2, 3.

*Exercise* 1. *Complete this table*

| Noun | Adjective |
|---|---|
| width | |
| breadth | |
| height | |

| Noun | Adjective |
|---|---|
| depth | |
| thickness | |

*Exercise 2. Describe the dimensions of these objects in as many ways as possible:*

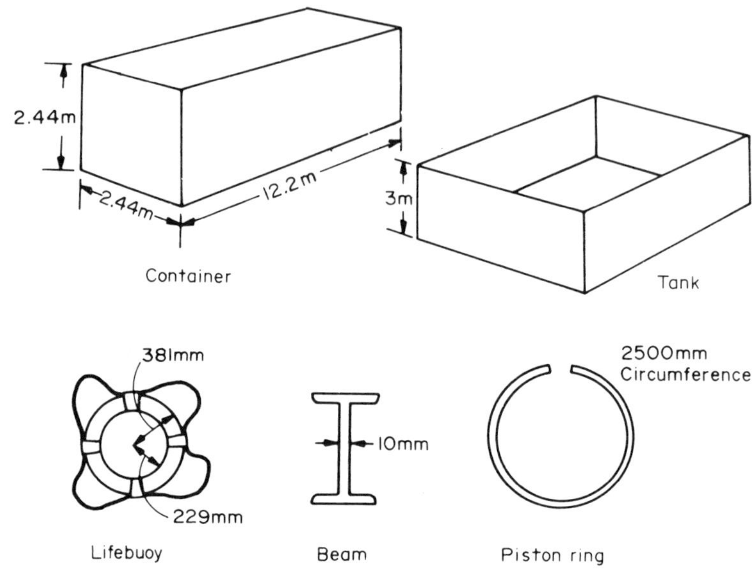


# GUIDED WRITING

## (A) Description of buoys

*Stage* 1. *Look at the notes below and study how they can be put together to form a description.*
(*Note:* all figures in this exercise should be considered as three-dimensional.)

**Porthand buoys**

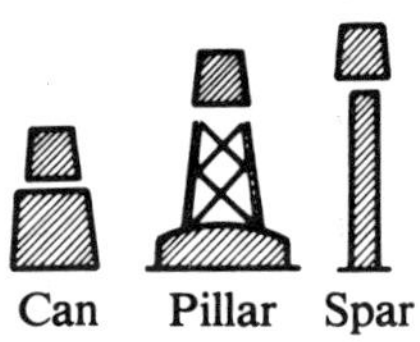


Colour: red
Shape: can, spar or pillar
Top mark: (if any) red can
Light: red, any rhythm

In the system of buoyage adopted by the International Association of Lighthouse Authorities (IALA), lateral marks consist of porthand buoys and starboard hand buoys.

Porthand buoys are red in colour. They are either can-shaped, pillar-shaped or spar-shaped. If they carry a top mark it is also red and shaped like a can. Lights, when fitted, are red and flash in any rhythm.

*Stage* 2. *Using the notes below and the description above as a guide, write a description of starboard hand buoys.*

**Starboard hand buoys**

Colour: green
Shape: cone, spar or pillar
Top mark: (if any) green cone
Light: green, any rhythm

*Stage* 3. *Now try and write a description of these special marks. They are all yellow in colour and so is their light when fitted. It flashes in any rhythm.*

**Special marks**

# NOTE-TAKING PRACTICE

## Abbreviations

When taking notes, the student tries to extract from a reading passage or lecture the important information and then put it on paper both briefly and coherently. To help him do this, a number of techniques must be developed. One technique is the use of abbreviations. Some of these are established, others can be invented. *Study the different forms of abbreviation below:*

1. Some established abbreviations:
   e.g. = for example     cf. = compare     no. = number
   i.e. = that is to say     etc. = and so on     NB = note well
2. Some formed from the first letter of a word or words:
   N = north     IALA = International Association of Lighthouse Authorities
   MV = motor vessel     IMO = International Maritime Organization
3. Some abbreviations formed from the first part of a word:
   approx. = approximately     poss. = possible, possibly
   auto. = automatic     gen. = general, generally
4. Some formed by omitting most vowels and some consonants, except the final one:
   mk = mark     govt = government     imprvd = improved
   ft = feet     dept = department     imprvmt = improvement
5. Note how the connectives studied in the grammar section and words and phrases having the same meaning can be abbreviated or simplified.

| | | |
|---|---|---|
| &, + | = | and, in addition to, also, moreover, not only . . . but also |
| or,/ | = | or, alternatively, either . . . or, on the other hand |
| but | = | but, whereas, yet, nevertheless, even so |

*Exercise* 1. *Write down these words and phrases in abbreviated form:*

(a) for example
(b) south
(c) either . . . or
(d) that is to say
(e) the United Nations
(f) number one
(g) compare
(h) in addition to
(i) note well
(j) alternatively
(k) north-east
(l) not only . . . but also
(m) and so on
(n) motor vessel
(o) yet

*Exercise* 2. *Use your imagination to write these down in abbreviated form:*

(a) possibly
(b) seconds
(c) special
(d) machinery
(e) optional
(f) amount
(g) porthand
(h) quick
(i) top mark
(j) group
(k) equipment
(l) introduction
(m) mathematics
(n) books
(o) accommodation

## **Lecture:** Buoyage

*You will now have a lecture describing more buoys used in the IALA Buoyage System 'A'. Listen to the lecture and complete the note outline below. The first part has been done for you.* The text of the lecture can be found in Appendix II. Do not look at the text until you have completed the exercise.

### *The IALA buoyage system 'A'*

1. Card. mks
   (a) N Card. Mk

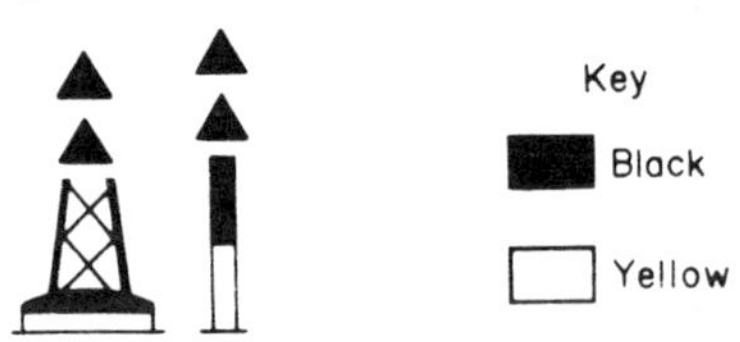


Col: black over yellow

Shpe: pillar or spar

Topmk: 2 black cones, pts upwds

Light: (when fitted) (i) col: white

(ii) rhythm: V Qk Fl

OR Qk Fl

(b) E Card. Mk

Col: ............................................

Shpe: ..........................................

Topmk: .......................................

Light: (when fitted) (i) col:................

(ii) rhythm:............

(c) ...... Card. Mk

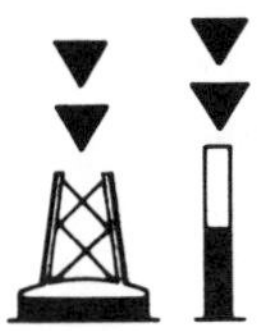

Col: ............................................

......:............................................

......:............................................

Light: (................) (i)......................

(ii) rhythm............

(d)..................

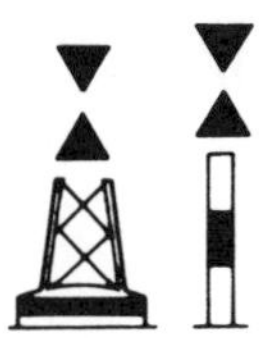

......:............................................

......:............................................

......:............................................

......: (................) (i)......................

(ii) ......................

2. Isolated Dngr Mk

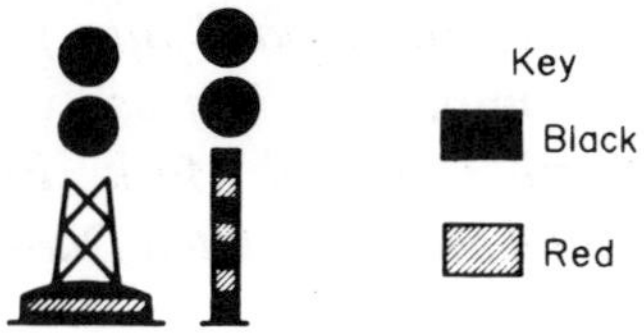


......:............................................

......:............................................

......:............................................

......: (................) (i)......................

(ii) ......................

3. ..................

......:..............................................

......:..............................................

......:..............................................

......: (................) (i)......................

(ii) .......................

*Now turn to Appendix II and read through the lecture checking your notes.*

(*Note:* **Isophase** – the period of light equals the period of darkness
**Occulting** – the period of light is longer than the period of darkness
**Flashing** – the period of light is shorter than the period of darkness.)

# GUIDED WRITING

## (B) Buoyage

*Use your notes to help you complete the description of the buoys used in the IALA Buoyage System 'A' which you started in Guided Writing (A).*

*Stage* 1. *Give your composition a title.*

*Stage* 2. *For your first two paragraphs, copy out the paragraphs in Guided Writing (A), Stage* 1.

*Stage* 3. *For your third paragraph add your description of starboard hand buoys, Stage* 2.

*Stage* 4. *On a new line, copy the sentence: 'Cardinal marks are divided into four groups corresponding to the four points of the compass'. Then use your notes to help you write a paragraph on a North Cardinal Mark, East Cardinal Mark, South Cardinal Mark, West Cardinal Mark, Isolated Danger Mark and Safe Water Marks.*

*Stage 5. Add on your description of special marks from Guided Writing (A), Stage* 3.

*Now read through your description carefully to check for spelling mistakes and errors.*

# UNIT II

# SHIP TYPES

## READING COMPREHENSION

### (A) Types of merchant ship (version 1)

*The following passage is about different types of merchant ship. Before you read the passage, write down on a piece of paper the names of any type of merchant ship you know. Try to divide your ships into groups based on the type of cargo they carry (e.g. passenger ships, liquid cargo ships). Now read the passage.*

[1]Merchant ships can be classified according to what they carry. [2]Most are designed to carry cargo, but a few still carry passengers.

[3]Cargo ships can be divided into two basic types. [4]One type carries dry cargo, the other carries liquid cargo. [5]Multi-deck vessels are a traditional type of dry cargo ship. [6]Their holds are divided horizontally by one or two 'tween decks. [7]Dry bulk cargo is carried in bulk carriers. [8]These do not have 'tween decks. [9]Container ships are the most modern type of dry cargo carrier. [10]They carry containers of standard dimensions. [11]Fruit, meat and dairy produce are carried in refrigerated ships. [12]Oil tankers are the most common type of liquid cargo carrier. [13]They are often very large. [14]Two other types of liquid bulk carrier of growing importance are liquefied natural gas (LNG) carriers and chemical carriers.

[15]In comparison with cargo vessels, passenger ships are fewer in number and type. [16]Passenger liners are the traditional type of passenger ship. [17]Nowadays their number has been greatly reduced. [18]Cruise ships are another type of passenger vessel. [19]These are often converted passenger liners. [20]Ferries are the most common type of passenger vessel. [21]Many of them are also designed to carry vehicles.

Were any of the ships that you had thought of mentioned here?

A summary of a reading passage can sometimes be made in the form of a tree diagram. *Complete the one below using the information from the reading passage.*

A general cargo carrier (*Unit II*)

A dry bulk carrier (*Unit II*)

A container ship (*Unit II*)

A liquid bulk carrier (*Unit II*)

A ferry (*Unit II*)

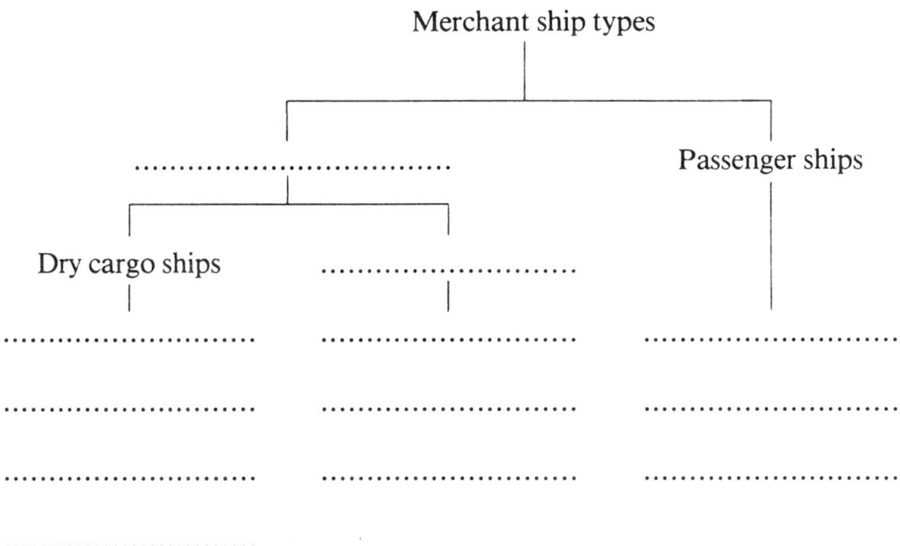


*Use your diagram and the passage to complete these statements:*

(a) Merchant ships can be designed as.........................................................

(b) Cargo ships can be divided into...............................................................

(c) Dry cargo ships include.............................................................................

(d) ..................................................are examples of liquid cargo carriers.

(e) Three types of passenger ship are..........................................................

*Now put these sentences together to form a paragraph. The paragraph will also be a summary of the reading passage.*

# GRAMMAR

## (A) Articles

### *(i) The, a, an, zero article*

Nouns in English can be preceded by the definite article (**the**) or by the indefinite article (**a, an**) or by no article at all.

The definite article (**the**) is used when the noun (singular or plural, countable or uncountable) being referred to has a particular rather than a general reference, that is, when we can identify what is being referred to.

Here are three situations when we can identify what is being referred to:

1. When the noun has been mentioned already
   e.g. Colliers are designed to carry *coal.* The coal is carried in bulk.

2. When words following the noun define which particular one it is
   e.g. Ships of all types use this port; the ship *over there* is an LNG carrier.
   (*Note:* The phrase 'over there' tells us which ship is being referred to, but 'of all types' does not.)

3. When there is only one of the noun in the world, or only one in the context being referred to
   e.g. the sun, the equator (with reference to the earth)
   the bridge, the Captain (with reference to a ship)

This quality of uniqueness is also present with superlatives
e.g. the largest ship, the most important port.

The indefinite article (**a, an**) is used in front of singular nouns when they are used as countable nouns and when the reference is general rather than particular, that is, when we do not say which example of the noun is being referred to:
e.g. There is a ship in the port.

When the noun is plural, or when the noun is used as an uncountable noun, no article is used.
e.g. There were cadets on board.
There was oil floating on the water.

When we refer to a class of objects rather than to examples of that class, the distinction between definite and indefinite use disappears, and there are three possible ways of referring to countable nouns:

e.g. The container ship is designed to carry containers.
A container ship is designed to carry containers.
Container ships are designed to carry containers.

(Here all three sentences mean: All container ships are designed to carry containers.)
**but** there is only one way of referring to uncountable nouns:

e.g. *Oil* is carried in tankers.

*Exercise* 1. *Write out the paragraph below using* ***a, an, the*** *or no article as appropriate:*

.......... largest type of cargo ship is .......... tanker. .......... tankers are designed to carry .......... liquid cargo such as .......... oil. .......... cargo is pumped directly into .......... holds by .......... powerful pumps. .......... holds are constructed as .......... tanks. .......... tanks are sub-divided into .......... central tank, two wing tanks and .......... expansion tank. .......... expansion tank allows .......... oil to expand in hot weather. .......... bridge superstructure and .......... engine room are situated aft to leave more room for .......... cargo. .......... bridge is connected to .......... forecastle by .......... catwalk. .......... tankers which are over 500,000 dwts are known as .......... ultra large crude carriers (ULCCs).

*Exercise* 2. *Add these words and phrases in turn to the basic sentence in the appropriate place. Put in* ***the*** *only where necessary. Study the examples.*

**Basic sentence:** Ships have crossed the Atlantic.

*Addition:* many
→ Many ships have crossed the Atlantic.
*Addition:* which we saw yesterday
→ The ships which we saw yesterday have crossed the Atlantic.

*Now add these in turn:*

(a) of all types
(b) largest
(c) merchant
(d) which arrived today
(e) loading in the port
(f) most

## (B) Logical connectives (ii)

Here are some more connecting words for joining statements: **because, therefore, however.**

*Study how they are used in these examples:*

1. **Because** gives the reason or cause
   - (a) Multi-deck vessels have 'tween decks.
   - (b) 'Tween decks help stowage.
   - (a) + (b) Multi-deck vessels have 'tween decks **because** these help stowage.

2. **Therefore** expresses consequence or result
   - (a) Ships are designed for many purposes.
   - (b) Their type and size vary considerably.
   - (a) + (b) Ships are designed for many purposes, **therefore** their type and size vary considerably.

3. **However** introduces a qualification or concession
   - (a) Passenger liners carry passengers.
   - (b) Some carry a large amount of cargo as well.
   - (a) + (b) Passenger liners carry passengers; **however,** some carry a large amount of cargo as well.

*Exercise* 1. *Now join these pairs of sentences using* ***because, therefore, however,*** *as appropriate:*

(a) Multi-deck vessels usually carry general cargo.
Some carry containers as well.

(b) Passenger liners have high superstructures.
They need a large number of decks.

(c) Many ferries are designed to carry vehicles.
They have doors at the bows or stern.

(d) Cargo ships are usually designed to carry dry or liquid cargo.
OBO (oil, bulk ore) ships are designed to carry both.

(e) Bulk carriers carry large quantities of loose cargo.
They have large unobstructed holds.

(f) Passenger liners often operate as cruise ships for part of the year.
There is not always enough business for them on liner routes.

Here are some more connecting words: **although, consequently, as.**
*Study the examples and decide if they introduce a clause of (1) reason or cause, (2) consequence or result, (3) qualification or concession.*

(a) Cargo liners sail on fixed routes and keep to a timetable, **consequently** some are designed to carry a few passengers.

(b) Bulk carriers do not usually carry derricks **as** loading and unloading is done by special cranes.
(c) **Although** tankers sail on fixed routes, they do not carry passengers.

*Your answers should have been:*
(a) consequence or result
(b) reason or cause
(c) qualification or concession

*Exercise* 2. *Rewrite the sentences in Exercise* 1 *above using* ***as, consequently, although,*** *as appropriate.*

# READING COMPREHENSION

## (B) Types of merchant ship (version 2)

*Read through this development of the reading passage in Reading Comprehension (A) and circle the words and phrases which the words underlined refer to. Note that when a singular noun is used with the definite article to refer to a class of objects, it is usually followed by a plural pronoun.*

[1]Merchant ships can be classified according to what they carry. [2]Most are designed to carry cargo, but a few still carry passengers.

[3]Cargo ships can be divided into two basic types. [4]One type carries dry cargo, the other carries liquid cargo; however, an OBO ship is designed to carry both. [5]A traditional dry cargo ship is the multi-deck vessel. [6]Her holds are divided horizontally by one or two 'tween decks, because these make stowage of individual packages easier. [7]Dry bulk cargo is carried in bulk carriers. [8]These do not have 'tween decks as cargo is carried loose. [9]The most modern type of dry cargo carrier is the container ship. [10]They carry containers of standard dimensions, consequently stowage is easier. [11]Fruit, meat and dairy produce are carried in refrigerated ships. [12]Oil tankers are the most common type of liquid cargo carrier. [13]They are often very large, because huge quantities of oil need to be transported and one large vessel is more economical to operate than two smaller ones. [14]Two other types of liquid bulk carrier of growing importance are the liquefied natural gas (LNG) carrier and the chemical carrier, although chemicals can also be carried in drums in general cargo ships.

[15]In comparison with cargo vessels, passenger ships are fewer in number and type. [16]The traditional passenger ship is the passenger liner; however, many carry cargo as well. [17]Nowadays their number has been greatly reduced, because of

competition from air transport. [18]Another type of passenger vessel is the cruise ship. [19]These are similar in appearance to passenger liners. [20]The most common type of passenger vessel is the ferry. [21]Many of them are also designed to carry vehicles, therefore these have doors at the stern or bows.

*Using the information in the passage and the connecting words studied in Grammar (B), carry out the following instructions:*

(a) add a **qualification** to this statement:
Cargo ships carry either liquid or dry cargo.
(b) add a **reason** to this statement:
Multi-deck vessels have 'tween decks.
(c) add a **consequence** to this statement:
Containers are of standard dimensions.
(d) add **two reasons** to this statement:
Oil tankers are often very large.
(e) add a **qualification** to this statement:
Chemicals are carried in chemical carriers.
(f) add a **qualification** to this statement:
Passenger liners are designed to carry passengers.
(g) add a **consequence** to this statement:
Many ferries carry vehicles.

When we qualify a statement we often add a reason.
*Use your imagination and knowledge to answer these questions:*

(a) Why are OBO ships designed to carry both liquid and dry cargo?
(b) How do 'tween decks help stowage of cargo?
(c) Why is it more economical to run one large tanker than two smaller ones?
(d) Why do general cargo ships sometimes carry chemicals?
(e) Why do passenger liners also carry cargo?

# APPLIED TERMINOLOGY

## (A) Terms relating to a ship's hull

The main body of a ship is called the **hull.**

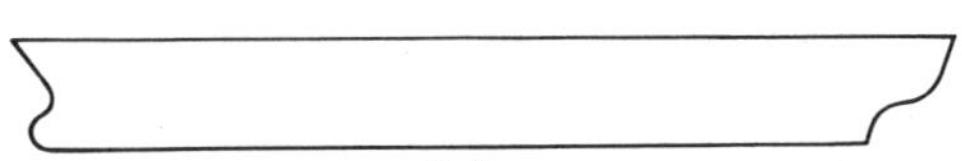

Hull

The hull is divided into three parts:

The foremost part is called the **bow** and the rearmost part is called the **stern**.

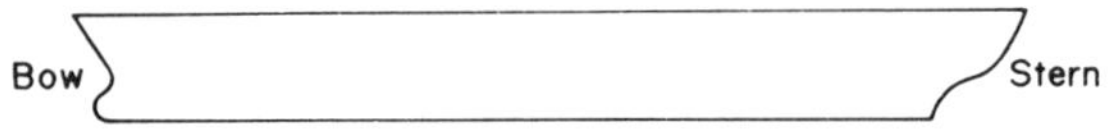


*Exercise* 1 *Cover the diagrams above and label and diagram below where indicated.*

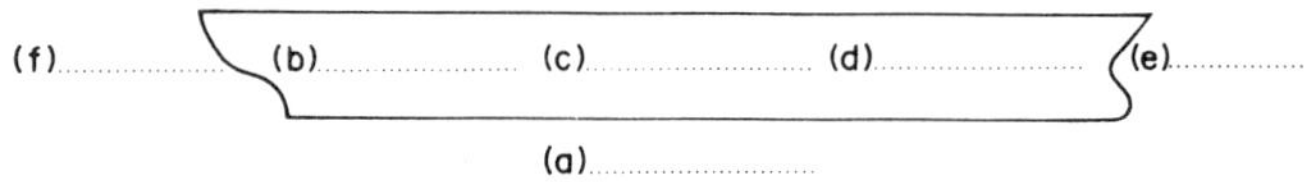


## Terms relating to the hull surface

The hull surface may be referred to using the following terms.

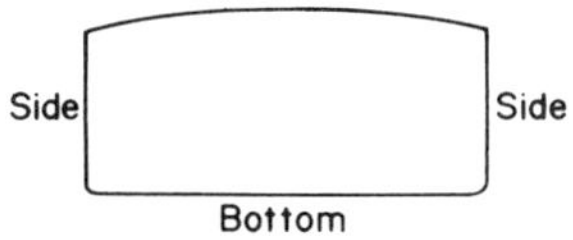


(*Note:* When standing in a ship and facing the bow, the left-hand side is called the **port side** and the right-hand side is called the **starboard side**. These can be added to the above terms.

e.g. the port bow, the starboard quarter, the port side amidships.)

The side of the hull can be referred to more accurately by using the side, the part and the waterline.

e.g. the ship was hit on the port bow one metre below the waterline.

*Exercise* 2. *Using the table to help you, use your imagination to complete the sentences below. The first has been done for you.*

| | | | | |
|---|---|---|---|---|
| port<br>starboard | bow<br>side amidships<br>quarter | .......... inches<br>.......... feet<br>.......... metres | above<br>below | the waterline |

(a) The hull was damaged on the port bow three feet above the waterline.

(b) The ship was holed..............................................................

(c) The stabilizers were fitted......................................................

(d) The sailors painted the hull....................................................

(e) There was thick armour plating................................................

## (B) Measurement

### *(ii) Ship's dimensions*

*Study the diagrams below showing some important ship dimensions:*

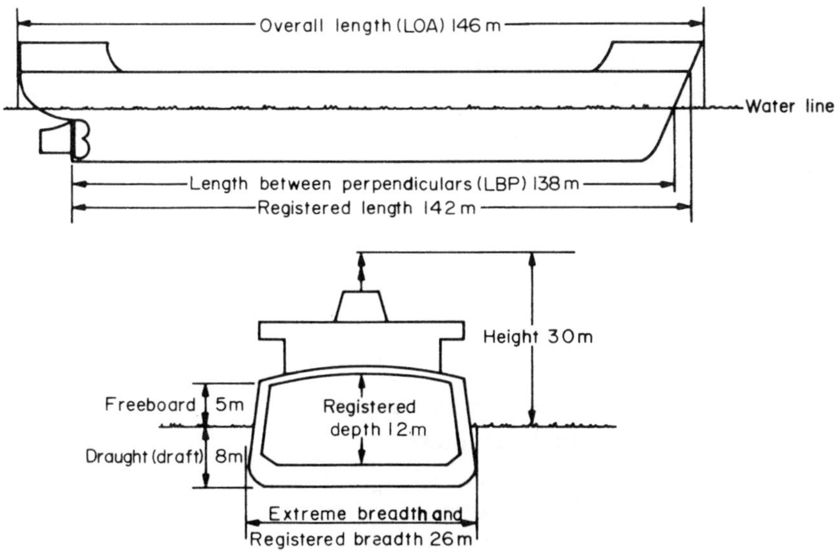


(*Note:* **Draught (draft), freeboard, height** depend on load carried. **Draught** may be expressed verbally.

e.g. The ship draws 8 metres.)

We can describe the dimensions of a ship using the following patterns:

1. The ship's overall length (LOA) is 146 metres.
2. The overall length (LOA) of the ship is 146 metres.
3. The ship has an overall length (LOA) of 146 metres.

*Exercise 1.* *Now look at the diagrams of the ship again and make similar statements about her other dimensions.*

You may be asked the following questions over the radio:
'What is your draught (forward/aft)/freeboard/height?'
To which you reply:
'My draught (forward/aft)/freeboard/height is . . .'

# GUIDED WRITING

## (A) Description of different types of warship

*Write a brief description of some of the different types of warship.*

*Stage* 1. *Combine the sentences in each group below to form short paragraphs. The first sentence in each group is the main statement and stays the same. The other two pairs can be combined to form two sentences using any of the connecting words studied so far.*

In the past the largest warships were battleships.
They were designed for heavy bombardment.
They carried up to sixteen-inch guns.
Their speed was relatively slow.
They were large and had heavy armour plating.

Nowadays aircraft carriers are the largest warships.
Most carry aircraft and helicopters.
A few carry helicopters only.
They have a narrow bridge superstructure.
This leaves more room for the flight deck.

Another type of warship is the cruiser.
These are smaller than battleships.
They are larger than destroyers.
They are designed to combine fire-power with speed.
They carry medium-sized guns and missiles.

*Stage* 2. *In the next two groups the order of the sentences has been confused. Choose the main statement and then sort out the other sentences joining them together as appropriate.*

These are expensive to run.
They are built for speed and manoeuvrability.
They may have diesel engines for cruising.
They are often powered by gas turbines.
Patrol boats vary greatly in design.

It carries guided missiles in addition to conventional guns.
Many frigates have an anti-submarine role.
Destroyers and frigates are designed for escort duties.
They carry weapons for destroying submarines.
The modern destroyer is taking over the role of the cruiser.

A modern destroyer (*Unit II*)

*Stage* 3. *Now try and write a brief paragraph on submarines. Use a reference book if necessary.*

*Stage* 4. In order to form a composition, paragraphs are put together in a natural and logical order.

*What sort of order do you think the first three paragraphs are in? Arrange the other three paragraphs in the same way.*

*Put all the paragraphs together to form a description of some of the different types of warship.*

A nuclear submarine (*Unit II*)

# NOTE-TAKING PRACTICE

## Symbols, omissions

Symbols are also used to cut down the amount of writing needed in note-taking. Many of these are used in mathematics.

*Study these symbols and their meanings.*

= : is, are, has, have, equals

∴ : therefore, consequently, thus, so

∵ : because, as, since.

Note also how one abbreviation can be used for words of similar meaning:

**but**: however, (al)though.

The words which are written down in note-taking, either in full or in abbreviated form, are only those words which carry information that is relevant. All other words can be omitted. The groups of words which we tend to omit when taking notes are listed below:

| | |
|---|---|
| 1. Auxiliaries | **be** (am, is, are, was, were), **have** (have, has, had), **do** (do, does, did) |
| 2. Articles | the, an, a, some |
| 3. Pronouns | he, him, his, himself etc., that, which, etc. |

4. Prepositions of time, direction, place, and when attached to verbs, adjectives and nouns.

(*Note:* The above list is only a guide and the note-taker must make up his own mind as to what can be left out either because it is not important or because it can be added when the notes are expanded.)

*Before listening to the lecture, try doing Exercises 1 and 2.*

*Exercise* 1. *Expand these notes into full sentences:*

(a) Tugs – 4 basic types (i) river, (ii) coastal, (iii) harbr, (iv) ocean-going.
(b) Dredgers remove sand & mud fro chnls & hrbrs.
(c) Icebreakers import ∵ N ports & chnls frze winter.
(d) Lightships look like ord ships BUT no eng ∵ towd postn & anchrd.
(e) Lifeboats – many types. In UK manned by vols & supp by voltry dons.
(f) Pilot launches = mtr boats for transport pilots to & fro ships.

*Exercise* 2. *Try and reduce these sentences to note form:*

(a) Tugs must be stable in all conditions, manoeuvrable and have sufficient power.
(b) Dredgers are of three main types: they can be either bucket dredgers, suction dredgers or grab dredgers.
(c) Ships must use these ports all the year round, therefore it is necessary to keep them open.
(d) Lightships not only have a light, but also a foghorn, a radio beacon and meteorological equipment.
(e) Lifeboats must be strong, stable and manoeuvrable and their crew must be well trained.
(f) Pilot launches must be seaworthy as pilots go out in all weathers.

## **Lecture:** Special duty vessels

*Listen to the lecture on some of the special duty vessels that you can expect to see around a port and complete the outline below using note-form throughout.*

Tugs – .......... basic types (i).........., (ii) harbr, (iii).............., (iv)..............-going

..........impt reqmts (i).........., (ii) manoeuvrable, (iii)..............

Icebreakers – import...........N ports........................frze..............................

shps MUST .................... all yr .................... open

have (i)........................, (ii)........................

Dredgers – remove ..........&..........fro chnls ..................................................

.......... main types (i)...................., (ii)................, (iii)..................

A dredger (*Unit II*)

Lightships – look like.................................................. towd.........................

.......... anchrd.

have light & ......................, ..........................., ...........................

crew = ..............................

Lifeboats – ............................. In..............manned by..............................

............................. by ..............................dons

.................... (i)...................., (ii)..................., (iii)...................

.......................................................................................

Pilot launches – .........................................................................

.........................................................................

.........................................................................

.........................................................................

*Now look at Guided Writing (B) and do the exercise there.*

# GUIDED WRITING

## (B) Special duty vessels

*Use your notes to help you write six paragraphs, one about each type of ship described in the lecture, i.e. tugs, dredgers, icebreakers, lightships, lifeboats and pilot launches.*

*When you have finished your descriptions, check them against the lecture in Appendix II. Were your notes adequate?*

# UNIT III

# SHIP CONSTRUCTION

## READING COMPREHENSION

### (A) Building ships

A reading passage is often divided into paragraphs. Each paragraph is made up of a number of sentences. These sentences are usually about the same topic. The topic of a paragraph may be summarized by a sentence or part of a sentence anywhere in the paragraph. By picking out the topic sentence, it becomes easier to understand the paragraph.

*Read the following passage about building ships and underline the topic sentence in each paragraph. The first one has been done for you. Do not stop for very long on words you do not understand.*

[1]Ships cost a lot of money to build. [2]A general cargo vessel costs several million pounds and a giant tanker can cost over £40 million pounds. [3]One reason for this is the high cost of steel and other materials used in shipbuilding. [4]Another reason is the high cost of labour.

[5]A modern shipyard is designed for building ships as cheaply and quickly as possible. [6]Many of the old processes have disappeared or been combined into one fully mechanized process. [7]Machines are now used instead of men. [8]Today, ships can be built in about sixteen months and costs can be kept to a minimum.

[9]Who designs ships? [10]Ships are designed by naval architects. [11]The largest shipping companies have their own naval architects. [12]In Europe and Japan, shipyards employ naval architects to design a ship for a customer, or offer basic designs which can be varied to suit the customer's needs. [13]Shipowners may also go to independent firms of shipping consultants and ask their naval architects to design a ship for them.

[14]When shipowners decide to order a new ship, they tell the naval architect the cargo they want the ship to carry. [15]They also tell him what routes the ship will ply and the desired speed. [16]They put limits on the ship's dimensions and on the price that they are prepared to pay. [17]The ship must also comply with the rules of the

classification society and international regulations. [18]Economic, engineering and safety factors all govern the design of a ship.

You should have underlined sentences 5, 10 and 18.

*Now read the passage through more carefully. Notice how words from everyday English often take on a slightly different meaning in more scientific English. For example, 'high' usually refers to height; what does it mean in sentences 3 and 4? An architect usually designs houses; what does a naval architect design, sentence 10? Governments govern people; what does 'govern' mean in sentence 18?*

*Also work out the relationship of the sentences within a paragraph to the topic sentence and to each other. For example, do they add examples (as in sentence 2) or reasons (as in sentences 3 and 4)?*

*When you have finished going through the passage, try placing the following words and phrases in the sentences indicated. It will sometimes be necessary to replace and re-order words.*

(a) for example (2)
(b) this is because of . . . and (3 and 4)
(c) with the result that . . . and (6 and 7)
(d) consequently (8)
(e) for instance (11)
(f) alternatively (12)
(g) in addition (13)
(h) not only . . . but (14 and 15)
(i) in addition (16)
(j) furthermore (17)
(k) to sum up (18)

# GRAMMAR

## (A) Passives

*Study these pairs of sentences:*

1. Naval architects design ships. **Active**
   Ships are designed by naval architects. **Passive**
2. Men built the tanker in six months. **Active**
   The tanker was built in six months. **Passive**
3. First, I measured the piece of metal. **Active**
   First, the piece of metal was measured. **Passive**

In each pair of sentences the meaning is similar, but the object in the active sentence has become the subject in the passive one.

The passive has a number of uses in English:

1. It allows the object of a sentence to be moved to the subject position. Here it becomes the main focus. In example 1, the active sentence would probably be used in a paragraph on naval architects, the passive version in one about ships. *Naval architects* now becomes the agent introduced with *by*.

2. It also allows the writer to avoid a subject like *men* in example 2. *Men* does not add any useful information and therefore can be left out of the passive version. The point of the sentence is not *who* built the tanker, but *how long* it took to build.

3. The passive also allows us to write in an impersonal style. This is useful in more scientific English, particularly for writing reports of experiments. Look at example 3. The passive not only makes it more objective, but also makes the reader focus on the action rather than the person doing the action. The person is unimportant here.

| The passive is formed by | **to be** (in the required tense) | + | **past participle** (of the main verb) |
|---|---|---|---|

Examples:

| | | |
|---|---|---|
| (a) are | built | (simple present) |
| (b) are being | built | (present continuous) |
| (c) was | built | (past simple) |
| (d) will be | built | (future) |
| (e) can be | built | (can) |
| (f) must be | built | (must) |
| (g) being | built | (present participle) |
| (h) to be | built | (infinitive) |

The pattern is the same for all tenses (a, b, c, d, etc.), for modals (e, f, etc.) and for gerund (g) and infinitive (h) forms.

*Exercise* 1. *Look again at the example pairs of sentences and then rewrite these sentences in the passive form:*

(a) Nowadays they make ships of steel.
(b) The Queen will launch the tanker tomorrow.
(c) They are painting the hull.
(d) The Royal Navy has ordered two new warships.
(e) They used to build ships of wood.
(f) You can join two pieces of metal together by welding.
(g) Next, we weighed the piece of metal.
(h) People must prepare the plates properly.
(i) They had to repair the bows.

*Exercise* 2. *Rewrite this paragraph putting most of the verbs into the passive to make it sound more formal:*

When we fit out a ship, we complete her. We put in the engines, if we have not already installed them. We finish off the superstructure and construct the accommodation for the crew. In addition, men erect masts and derricks and put various items of deck machinery in place. Outside companies usually do the electrical work, plumbing and any woodwork. Someone must also buy the furniture and fittings for all the saloons and cabins. All this work which we do after we have launched a vessel, we call 'fitting out'.

## (B) Time relaters

### *(i) Sequence*

When we want to show that processes or events happen one after the other, we use sequence words or time clauses.

1. Sequence words

Some common sequence words are: **first, then, next, after that, afterwards, later, eventually, finally.** These are usually put at the beginning of the process or event that they introduce:

e.g. **First**, I went to college . . . . **Then,** I went to sea. . . .

These events may be described in a series of sentences, or they may be linked by a semi-colon (;), or the connective **and.**

Except for **first** and **finally** they may be used in any order.

*Exercise* 1. *Study the sentences below which show the sequence of events in the building of a ship. Write them out in a paragraph using the above sequence words to introduce each stage in a sentence.*

1st – The plans are completed by the naval architects.
2nd – The plans are approved by the classification society.
3rd – The parts of the ship are prepared.
4th – The parts of the ship are put together.
5th – The ship is launched.
6th – The ship is fitted out and completed.
7th – The ship goes for sea trials.
8th – The ship is handed over to her new owners.

2. Time clauses

Time clauses can begin with **after** and **before.** *Study these examples, which show how the first two sentences in the above description can be joined together using time clauses:*

**After** the plans are completed by the naval architects, they are approved by the classification society.

**Before** the plans are approved by the classification society, they are completed by the naval architects.

This time we are joining the first two sentences to form one sentence; *the plans* in the time clause becomes *they* in the main clause.

*Exercise 2. Using the events in the previous exercise, complete these sentences:*

(a) After the parts of the ship are prepared,........................................

(b) ..............................................., she is fitted out and completed.

(c) ..........................................................., she goes for sea trials.

3. Reduced time clauses

When the subject of the time clause is the same as the subject of the main clause, the time clause can be reduced in the following way:

**After** the plans are completed by the naval architects, they are approved by the classification society.

⇒ **After** being completed by the naval architects, the plans are approved by the classification society.

*Exercise 3. Reduce the sentences that you have completed in Exercise 2 so that they are like the example above.*

# READING COMPREHENSION

## (B) Building ships (cont.)

So far we have seen that sentences are linked grammatically, for example a noun in one sentence becomes a pronoun when referred to again. Sentences are also linked by meaning. That is to say, words are often repeated to show the meaning link between sentences, though they may be in a different form. For example the past participle *completed* may appear in the next sentence as the noun *completion.* Also, in order to avoid using the same word twice, a writer will use a word or phrase with the same meaning, for example, *ship* may become *vessel.*

*Read through the following passage, which is a continuation of the reading passage on page 32. As you read, mark the passage to show the meaning links between sentences. These links may or may not be in consecutive sentences. The first is done for you.*

[1]The building of a ship follows a well-ordered sequence of events. [2]After the vessel has been ordered, the plans are completed in the drawing-office. [3]Next, the final plans must be approved by a classification society such as Lloyds Register of Shipping. [4]This is necessary if the owner wants his ship to be classed. [5]While the ship is being built, constant checks are made to make sure she is being built to the standards of the society. [6]Classification will show that the ship is seaworthy and able to carry the cargo she has been designed to carry.

[7]Nowadays a shipyard is organized so that each stage in the building of a ship is done in a continuous chain of shops. [8]Each shop is linked by conveyor rollers and moving cranes on rails. [9]First of all, steel plates and bars are taken from the stock-yard to the preparation shop. [10]Here they are cleaned by shot blasting. [11]Then, they are coated with a primer paint to prevent corrosion. [12]Later, they are cut and shaped automatically by machines. [13]Cutting is done by gas torches and shaping by giant presses. [14]After that, the pieces are welded together in prefabrication sheds to form sections. [15]Welding is now used instead of riveting for joining pieces of metal together. [16]Riveting uses more steel than welding and was therefore more expensive. [17]It also increases the weight of the ship without increasing the strength. [18]The prefabricated sections are then transferred to the building berth. [19]Eventually, they are lifted into position by giant cranes.

[20]When a ship is ready, she is launched. [21]Some ships are built on a slipway and slide into the water. [22]Others are built in a dry dock. [23]The dock is then flooded with water and the ship is floated out. [24]After being launched, she is towed to the fitting out basin by tugs and completed.

[25]A completed ship goes for sea trials before she is handed over to her new owners. [26]During these the ship and her equipment are thoroughly tested.

*Now use the information in the reading passage to write out the following sentences in the correct sequence giving an outline summary of the building of a ship:*

(a) The ship is launched.
(b) The ship and her equipment are tested at sea.
(c) The plans receive the approval of the classification society.
(d) The new owners take possession of their vessel.
(e) The order for the ship is placed.
(f) The parts of the ship are prepared.
(g) The sections are assembled on the building berth.
(h) The final plans are drawn up.
(i) The parts are welded together to form sections.
(j) The ship is fitted out.

The stockyard (*Unit III*)

In the preparation shop: a cutting machine (*Unit III*)

In the prefabrication shop: welding the sections (*Unit III*)

The building berth (*Unit III*)

# APPLIED TERMINOLOGY

## (A) Terms relating to the hull

The main part of a ship is the hull. This is the area between the **main deck,** the **sides** and the **bottom.**

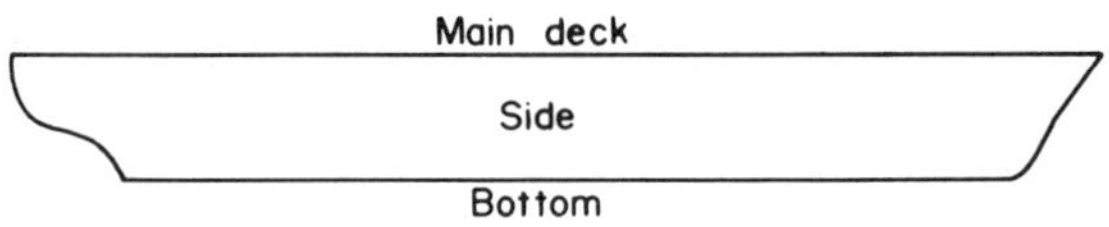


It is made up of **frames** covered with **plating.**

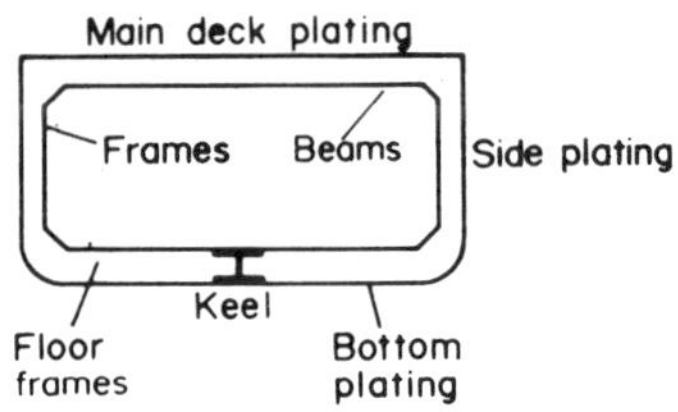


The hull is divided up into a number of watertight compartments by **decks** and **bulkheads.** Bulkheads are vertical steel walls going across the ship and along. Decks divide the hull horizontally. Those dividing up cargo spaces are known as **'tween decks.** The hull contains the **engine room,** cargo space and a number of tanks. In dry cargo ships the cargo space is divided into **holds,** in liquid cargo ships it is divided into **tanks.** At the fore end of the hull are the **fore peak tanks** and at the after end are the **after peak tanks.** They are used for fresh water and water ballast. The space between the holds and the bottom of the hull contains **double bottom tanks**. These are used for ballast water and fuel.

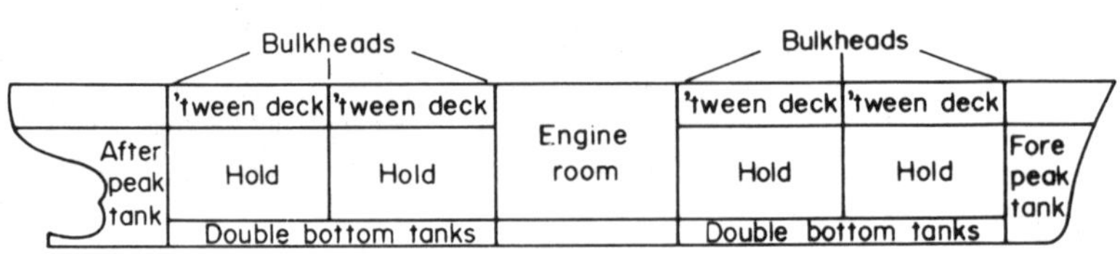

## Terms relating to superstructure

All permanent housing above the main deck is known as **superstructure**. The basic pattern of superstructure for a traditional dry cargo ship is shown below.

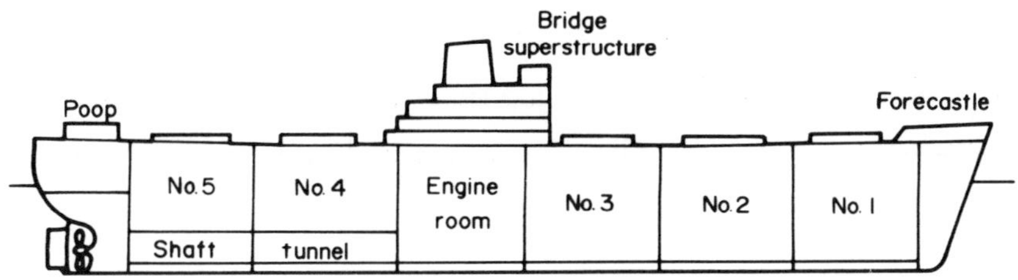


*Exercise 1. Complete this description of a modern cargo ship:*

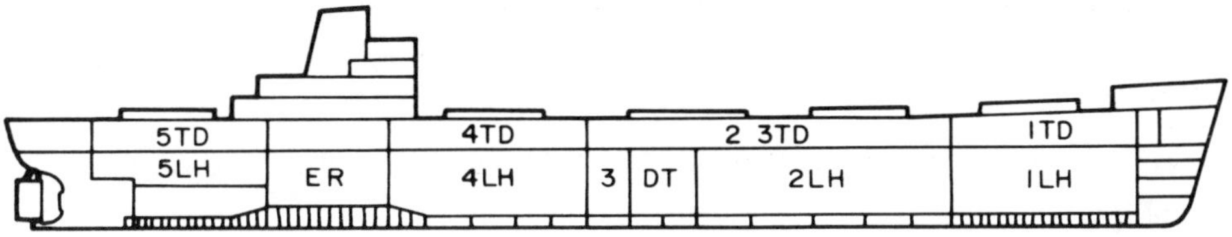


The diagram above shows the layout of a modern dry cargo ship. The hull is divided up into a number of watertight ............... by decks and steel ............... At the fore and after ends of the hull are the ............... tanks and the ............... tanks. The ............... is situated at the after end of the ship to leave more room for cargo. The cargo space is divided up into ............... These also have ............... Above the main deck is the ............... At the fore end is the ............... At the after end the ............... superstructure and the ............... are combined.

*Exercise 2. Now write a similar description of this tanker:*

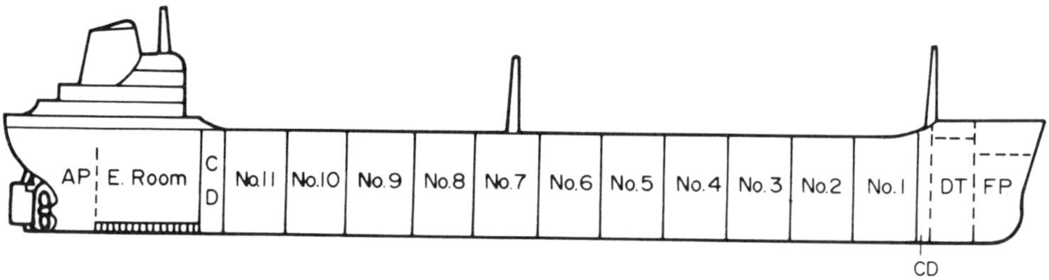


## (B) Measurement

### *(iii) Displacement*

From Archimedes' Principle we know that a ship when fully afloat displaces a weight of water equal to its own weight. Therefore a ship's weight is expressed in terms of displacement.

There are different displacements for different types of ship and condition:

for **warships**

| | |
|---|---|
| Displacement: | The weight of water displaced when loaded with fuel, water, stores, ammunition and with crew on board |
| Standard displacement: | The weight of water displaced when loaded with stores, ammunition and crew on board, but *without* fuel and reserve feed water |

for **merchant ships**

| | |
|---|---|
| Load displacement: | The weight of water displaced when loaded to her marks with cargo, stores, fresh water, fuel, water ballast, crew, passengers and baggage |
| Light displacement: (lightweight) | This is the weight of the hull of a ship and her machinery and spare parts, and with water in her boilers and condensers to working level plus lubricating oil and cooling water |

*Exercise 1. Label these disagrams with* ***displacement, standard displacement, load displacement, light displacement*** *(lightweight) as appropriate.*

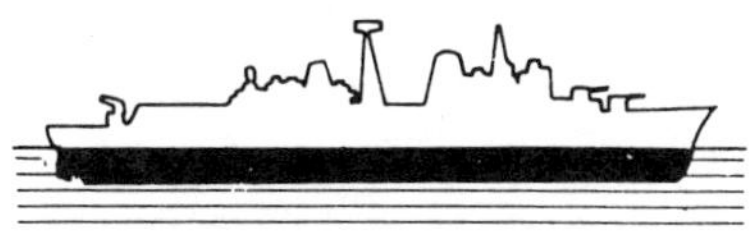

(a) ..........................

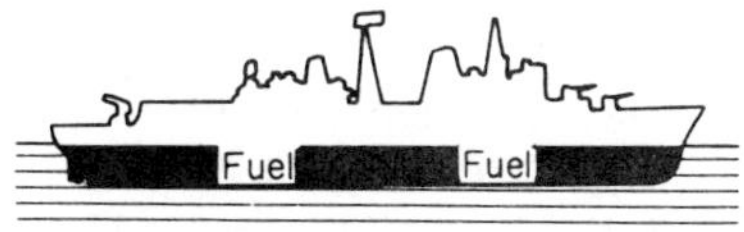


(b) ..........................

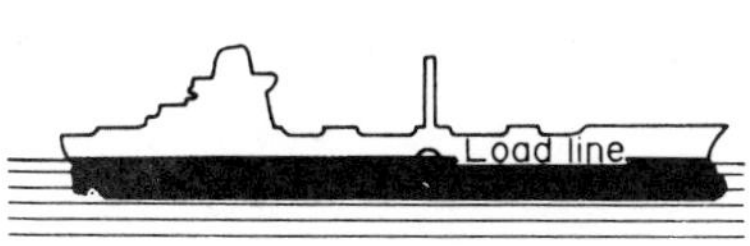


(c) ..........................

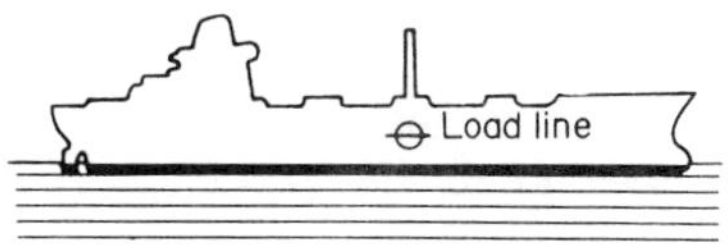


(d) ..........................

# GUIDED WRITING

## (A) Description of building a ship's hull

*Write a brief description of the sequence of events in the building of a ship's hull.*

*Stage* 1. *Study this flow chart, which shows each stage the plates and bars go through on their way to the building berth:*

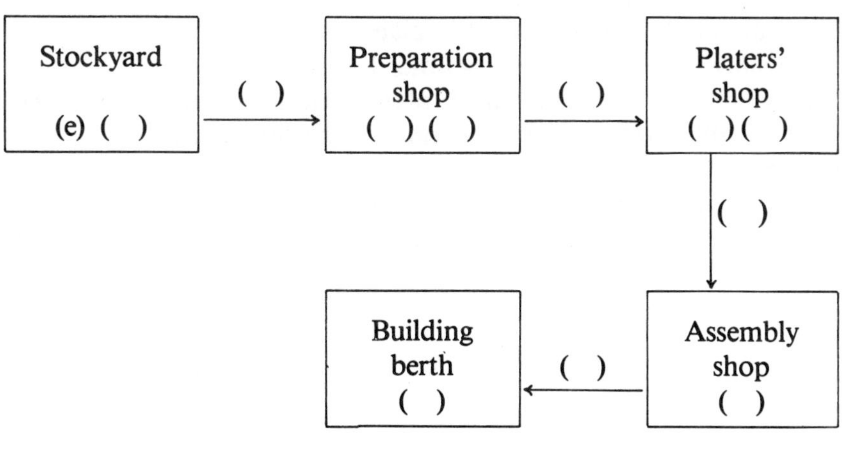

*Stage* 2. *Read through these sentences and put them into the correct sequence on the diagram above. The first (e) has been done for you.*

(a) The units are carried to the building berth by giant cranes.
(b) In the platers' shop they are cut to the correct size by gas torches.
(c) They are conveyed to the preparation shop by magnetic cranes.
(d) In the assembly shop they are joined together to form larger units.
(e) The steel plates and bars are off-loaded from lorries or railway wagons.
(f) At the building berth the units are welded together to form the hull.
(g) They are transferred to the platers' shop by conveyor rollers.
(h) In the preparation shop they are coated with primer paint to prevent corrosion.
(i) They are transferred to the assembly shop by cranes or conveyor rollers.
(j) In the platers' shop they are shaped by different presses and rollers.
(k) The steel plates and bars are stored in the stockyard ready for processing.
(l) In the preparation shop they are cleaned by shot blasting.

*Stage* 3. *Put the sentences together into a paragraph describing the building of a ship's hull. Introduce each stage with sequence words or time clauses.*

# NOTE-TAKING PRACTICE

## Ordering, layout

In a lecture, a lecturer may use the following words and phrases to introduce a sequence of events or a series of ideas:

**first (of all); firstly; at first; in the first place**

He will possibly introduce other events and ideas with:

**secondly, thirdly, etc.: then; next; after that; afterwards; later; eventually; subsequently; following; after; before; prior to; again; another**

He will perhaps finish with:

**last (of all); lastly; at last; ultimately; finally**

In note-taking all these can be abbreviated by using numbers and letters of the alphabet:

| | | | | |
|---|---|---|---|---|
| 1 | I | (i) | A | (a) |
| 2 | II | (ii) | B | (b) |
| 3 | III | (iii) | C | (c) |
| etc. | etc. | etc. | etc. | etc. |

When using these abbreviations the layout should be clear:

e.g. I ................................

(i) ...........................

II ..............................

(i) ...........................

(ii) ..........................

1. ................................

(a) ...........................

(b) ...........................

(i) ........................

2. ...............................

## **Lecture:** Improvements in cargo ship design

*Listen to the passage for Practice 1 on the improvements in the design of cargo ships and complete the first outline below, using notes:*

Imprvmts in ship design

1. Ships now faster

2. ...............................

3. ...............................

4. ...............................

*Now listen to the passage for Practice 2 and complete this outline (this time more information has been added):*

..................................................................

1. ...............................

∵ (a)...............................

2. ...............................

..... (a)...............................

3. Imprvd use of hull space

..... (a)...............................

4. ...............................

..... (a)...............................

*Now listen to the complete lecture on the improvements in the design of cargo ships and make your own notes. This time there is even more information and you will have to expand the basic outline.*

## GUIDED WRITING

### (B) Improvements in cargo ship design

*Without turning back to Appendix II use your notes to help you write a short description of some of the improvements in the design of ships. Your description should consist of a very short introduction and four short paragraphs.*

# UNIT IV

# MANNING

## READING COMPREHENSION

### (A) The organization of a ship's crew

A reading passage often has a diagram to go with it. This diagram should help you to understand the reading passage. It will give you some idea of what the passage, or part of the passage, is about. It may also give you some idea of the structure of the passage.

*Study this diagram. What do you think the passage is about? In what order do you think the writer will discuss the organization of a ship's officers and crew?*

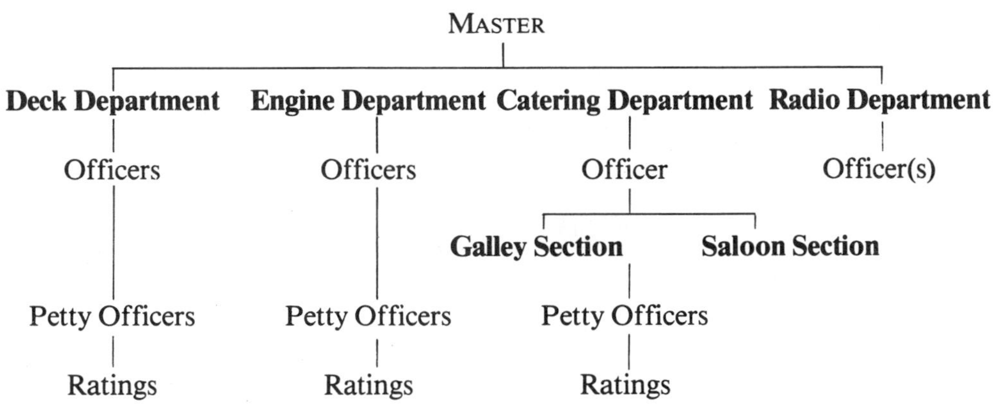


*Now read the passage*

The man in charge of a ship is the Master. He is responsible for the ship, her cargo and the safety of the crew. He must be well qualified and an experienced navigator.

Although his correct title is the Master, he is addressed as 'Captain'.

The organization of the crew of a cargo ship is changing, but it is still customary to find Deck, Engine, Catering and Radio Departments in ships of a reasonable size. Each department is made up of a varied number of officers, petty officers and ratings.

The Chief Officer, or First Mate as he is often called, is the Master's chief officer and head of the Deck Department. He is assisted by a Second Officer (Mate), a Third Officer (Mate), and sometimes a Fourth Officer (Mate). Several companies employ a First Officer as well as a Chief Officer. The Deck Department also includes a Boatswain (Bosun) and a Carpenter, both petty officers, and a number of ratings. These are made up of Able Seamen (AB), Ordinary Seamen (OS) and a middle grade known as Efficient Deck Hands (EDH). There are other grades of seamen. On some ships Navigating Cadets are carried for training purposes.

The Chief Engineer is head of the Engine Department. He is assisted by a Second, Third, Fourth and sometimes Fifth Engineer. An Electrical Officer may also be carried. The engine room petty officers are the Storekeeper and Donkeyman. On tankers there is also a Pumpman. He is also a petty officer. The engine room ratings are Firemen and Greasers. There may also be Engineer Cadets.

The Catering Department is under the Chief Steward. It is divided into a saloon and galley section. The former is headed by the Second Steward, the latter by the Ship's Cook. They are both usually petty officers. They are assisted by several stewards and cooks, and by a number of junior ratings.

The Radio Department often consists of only one man: the Radio Officer. On ships where continuous radio watches are kept there may be three radio officers: a Chief, Second and Third.

*Now use the information in the passage to expand the diagram at the beginning. Your diagram should show how each department is made up. If personnel exist only on some ships, put their names in brackets (   ).*

# GRAMMAR

## (A) Function

A person's function, or what he does, can be expressed in terms of his responsibility.

*Study these examples:*

1. The Master <u>is responsible for</u> the safety of the ship.
2. The safety of the ship is <u>the responsibility of</u> the Master.

*Note the addition here:*

3. The Master is responsible to the company for the safety of the ship.

*Exercise* 1. *Link the following (do not change their order), using whichever pattern above is appropriate:*

(a) Chief Officer – Master – the Deck Department.
(b) Third Officer – the life-saving equipment.
(c) The sounding of tanks and bilges – Carpenter.
(d) Radio Officer – radio communications.
(e) Chief Steward – Master – the Catering Department.
(f) The preparation of food – Ship's Cook.
(g) Chief Engineer – the efficient running of his department.
(h) The loading and unloading of oil – Pumpman.

The function of a thing, or what it is used for, can be expressed in a number of ways:

1. By using the phrase: **The function of . . . is to . . .**
   e.g. The function of a crane is to lift heavy objects.

2. By using the verb **to use + for -ing**
   e.g. A crane is used for lifting heavy objects.

3. By using a verb expressing the function.
   e.g. A crane lifts heavy objects.

4. By using a prepositional phrase introduced by **with**
   e.g. We lift heavy objects with a crane.

*Exercise* 2. *Rewrite the following sentences in the three alternative ways.*

(a) The function of a thermometer is to measure temperature.
(b) A fire extinguisher is used for putting out fires.
(c) A windlass raises and lowers the anchors.
(d) We measure time with a chronometer.

## (B) Time relaters

### *(ii) Simultaneous time*

When we want to show that one event takes place at the same time as another we can link these events using **when, while** and **as**.

1. **When** and **while**

**When** tells us what time two simultaneous actions happen. It also implies that the two events are completed, unless otherwise stated.

e.g. When the ship is launched, the crowds cheer.

**While** tells us that one action is happening at the same time as another. It is often used with continuous tenses.

e.g. While the men were preparing for the launch, the people started to arrive.

*Exercise* 1. *Fill in the blanks with* **when** *or* **while** *as appropriate:*

(a) .............. the Captain steps on board, he is saluted.

(b) .............. the main engine is being installed, the auxiliary machinery is fitted.

(c) Constant checks are made, .............. the ship is being built.

(d) .............. the surveyor inspects the lifeboats, he examines their equipment as well.

(e) .............. the cargo holds are being lined, the derricks are assembled.

(f) Different types of paint are used, .............. the ship is painted.

2. Reduced **when** and **while** clauses

Clauses introduced by **when** and **while** can be reduced in the same way as clauses beginning with **after** and **before**, that is, if the subject of the main clause is the same as the subject of the time clause.

e.g. While the ship is being built, she is constantly being tested.

⇒ While being built, the ship is constantly being tested.

*Exercise* 2. *Reduce the following sentences, if it is possible to do so:*

(a) When the ship undergoes speed trials, she produces her maximum speed.
(b) While the main engine is being installed, the auxiliary machinery is fitted.
(c) While the engines are running, they are carefully checked.
(d) When the surveyor inspects the lifeboats, he examines their equipment as well.

(e) When a ship is tested, many people watch her progress.
(f) While the ship covers the mile distance, she keeps a straight course.

### 3. **As**

When two actions are closely connected, **as** can be used instead of **while**
e.g. As the ship was leaving the harbour, she hit the jetty.

**As** is often used when the action in the time clause is the cause of the action in the main clause.
e.g. As the sun rose, the sky became lighter.
(*Note:* Clauses introduced by **as** cannot be reduced.)

# READING COMPREHENSION

## (B) The deck department

*Read through the passage below describing the work of the Deck Department. As you read, mark the passage to show the meaning links between the sentences. Remember that words often appear again in a different form (as a noun instead of a verb), or another word or phrase having the same meaning may be used. One link has been done for you.*

[1]The Deck Department is responsible for navigating the ship safely and economically from port to port. [2]The Master is an experienced navigator and usually works out the best course. [3]The Second Officer is responsible to the Master for keeping the ship on course and for looking after all the equipment used for navigation. [4]It is also the job of the Deck Department to see that the cargo is stowed properly in the holds and kept in good condition during the voyage. [5]The stowage of cargo is the responsibility of the Chief Officer. [6]He is helped by the Second and Third Officers. [7]In addition, when the ship is not fully loaded, the First Mate must see that the holds are cleaned and prepared for their next cargo. [8]In a tanker the cargo tanks are washed out during ballast passages and freed of gas. [9]At sea, much of the Deck Department's time is spent maintaining the ship and her equipment in good condition. [10]This means constant cleaning, painting and repair work. [11]This is done by ratings under the supervision of the Boatswain (Bosun). [12]A programme of maintenance for each day is worked out by the Chief Officer. [13]He also looks after the general day-to-day running of the department and deals with any problems.

[14]The Third Officer is in charge of the life-saving equipment. [15]The different appliances must be complete and in good working order.

[16]The Boatswain and the Carpenter are directly responsible to the Chief Officer. [17]The Bosun sees that his orders and those of other deck officers are carried out by the crew. [18]He is a man with a lot of knowledge and practical experience in seamanship. [19]The Carpenter is usually a qualified shipwright. [20]He no longer works only with wood as his name suggests. [21]His most important regular job is to sound the tanks and bilges in order to check the depths of liquid in them. [22]He also operates the windlass, when the anchors are being raised or lowered.

[23]The Deck Department is also responsible for keeping watches. An officer is always on watch on the bridge. [24]He is the Master's representative and answers to him for the safety of the ship during his watch. [25]In ships where a Chief Mate and a First Mate are carried, the First Mate is the watchkeeping officer.

*Now complete these sentences with one word from the passage:*

(a) The Second Officer is the .............. officer, therefore he must be good at calculating the ship's position and course.

(b) The careful .............. of cargo is important in order to keep it safe and easily unloaded.

(c) Cleaning, painting and repair work is known as ..............

(d) Life-saving equipment is the .............. of the Third Officer.

(e) On passenger liners there are always two officers on .............. on the bridge.

(f) It is the job of the Carpenter to .............. the tanks and bilges.

*Write out five true statements from this table:*

| | | | |
|---|---|---|---|
| The | Chief Officer<br>Second Officer<br>Third Officer<br>Boatswain<br>Carpenter | is responsible for | navigation<br>sounding tanks and bilges<br>supervizing the crew's work<br>cargo and maintenance<br>the life-saving equipment |

# APPLIED TERMINOLOGY

## (A) Terms relating to position in a ship

At the fore end of a ship is known as **forward**
At the after end of a ship is known as **aft**
At the midships part is known as **amidships**

e.g. The forecastle is situated forward.
The main engines in a tanker are aft.

**Port side** and **starboard side** can be added to these:

e.g. port side forward, starboard side amidships, port side aft.

The extreme fore end is known as **right forward**
The extreme after end is known as **right aft**

e.g. The ensign is right aft.

*Exercise* 1. *Draw a diagram of a ship's hull and label it to illustrate the meaning of the words described above. Your diagram should be the plan view of a hull, i.e. from above looking down.*

Here are some more terms relating to position:

In front of is known as **before** or **forward of**
Behind is known as **abaft** or **aft of**

Across the ship from side to ship is **athwartships**
Along the length of the ship from stem to stern is **fore and aft**

e.g. Number 1 hold is forward of Number 2 hold.
The funnel is always abaft the bridge.
Life rafts can be stowed athwartships or fore and aft.

*Now study this diagram of a traditional general cargo ship and read the description below:*

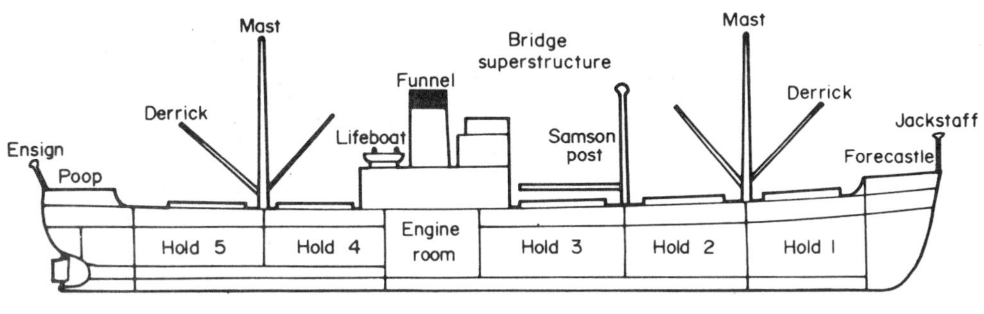

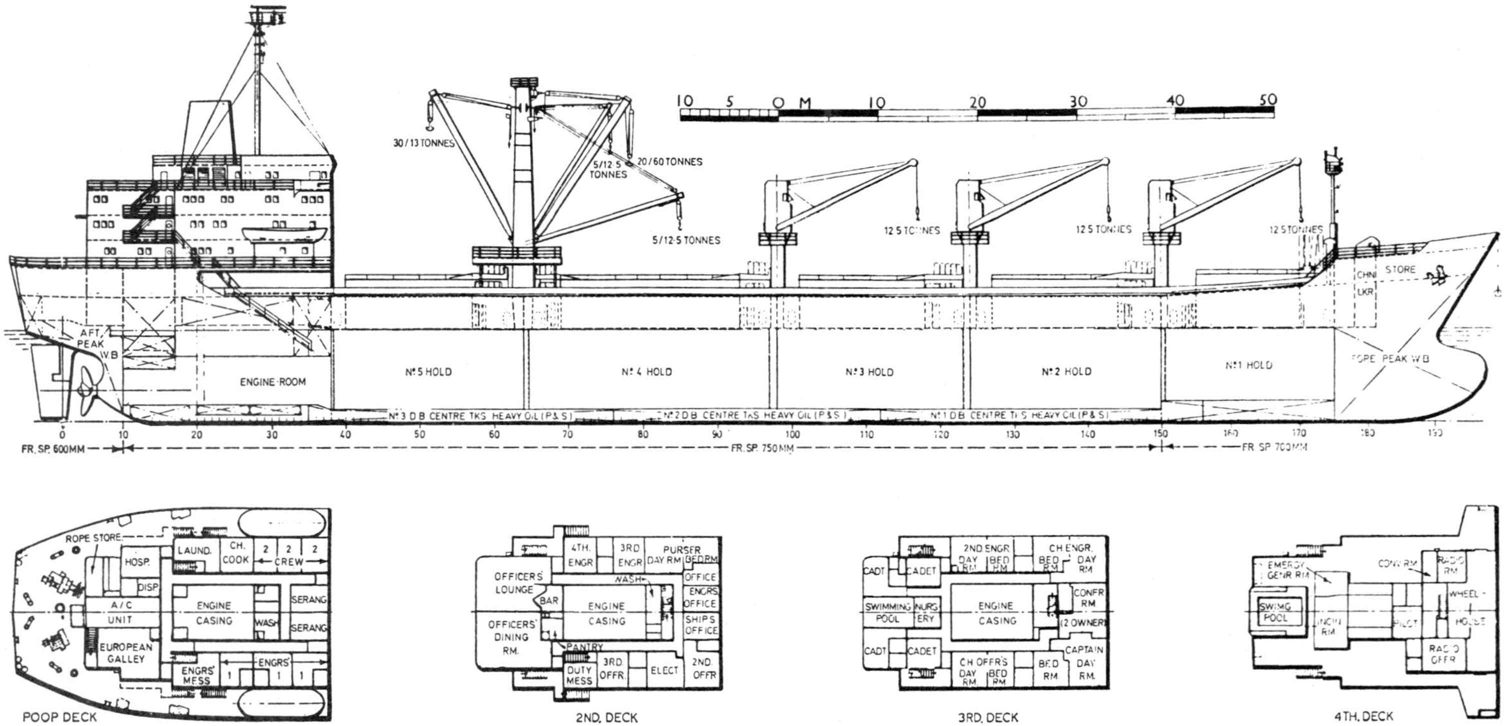

10 5 0 M 10 20 30 40 50
30/13 TONNES
20/60 TONNES
5/12·5 TONNES
5/12·5 TONNES
12·5 TONNES
12·5 TONNES
12·5 TONNES
CHN LKR
STORE
FORE PEAK W.B
AFT PEAK W.B
ENGINE ROOM
Nº5 HOLD
Nº4 HOLD
Nº3 HOLD
Nº2 HOLD
Nº1 HOLD
Nº3 D.B. CENTRE TKS HEAVY OIL (P & S)
Nº2 D.B. CENTRE TKS HEAVY OIL (P & S)
Nº1 D.B. CENTRE TKS HEAVY OIL (P & S)
0 10 20 30 40 50 60 70 80 90 100 110 120 130 140 150 160 170 180 190
FR. SP. 600MM
FR. SP. 750MM
FR. SP. 700MM
ROPE STORE
HOSP.
DISP.
A/C UNIT
EUROPEAN GALLEY
LAUND.
CH. COOK
CREW
ENGINE CASING
WASH
SERANG
SERANG
ENGRS' MESS
ENGRS'
POOP DECK
OFFICERS' LOUNGE
OFFICERS' DINING RM.
BAR
PANTRY
DUTY MESS
4TH. ENGR
3RD ENGR
PURSER DAY RM
BEDRM
OFFICE
ENGRS OFFICE
SHIPS OFFICE
WASH
ENGINE CASING
3RD. OFFR.
ELECT
2ND. OFFR
2ND. DECK
CADT
CADET
SWIMMING POOL
NURSERY
2ND ENGR. DAY RM
BED RM
CH ENGR. BED RM
DAY RM
ENGINE CASING
CONFR RM
(2 OWNER)
CH OFFR'S DAY RM
BED RM
CAPTAIN BED RM.
DAY RM.
3RD. DECK
EMERGY GENR RM
SWIMG POOL
CONV RM
RADIO RM
WHEEL HOUSE
PILOT
RADIO OFFR
4TH. DECK

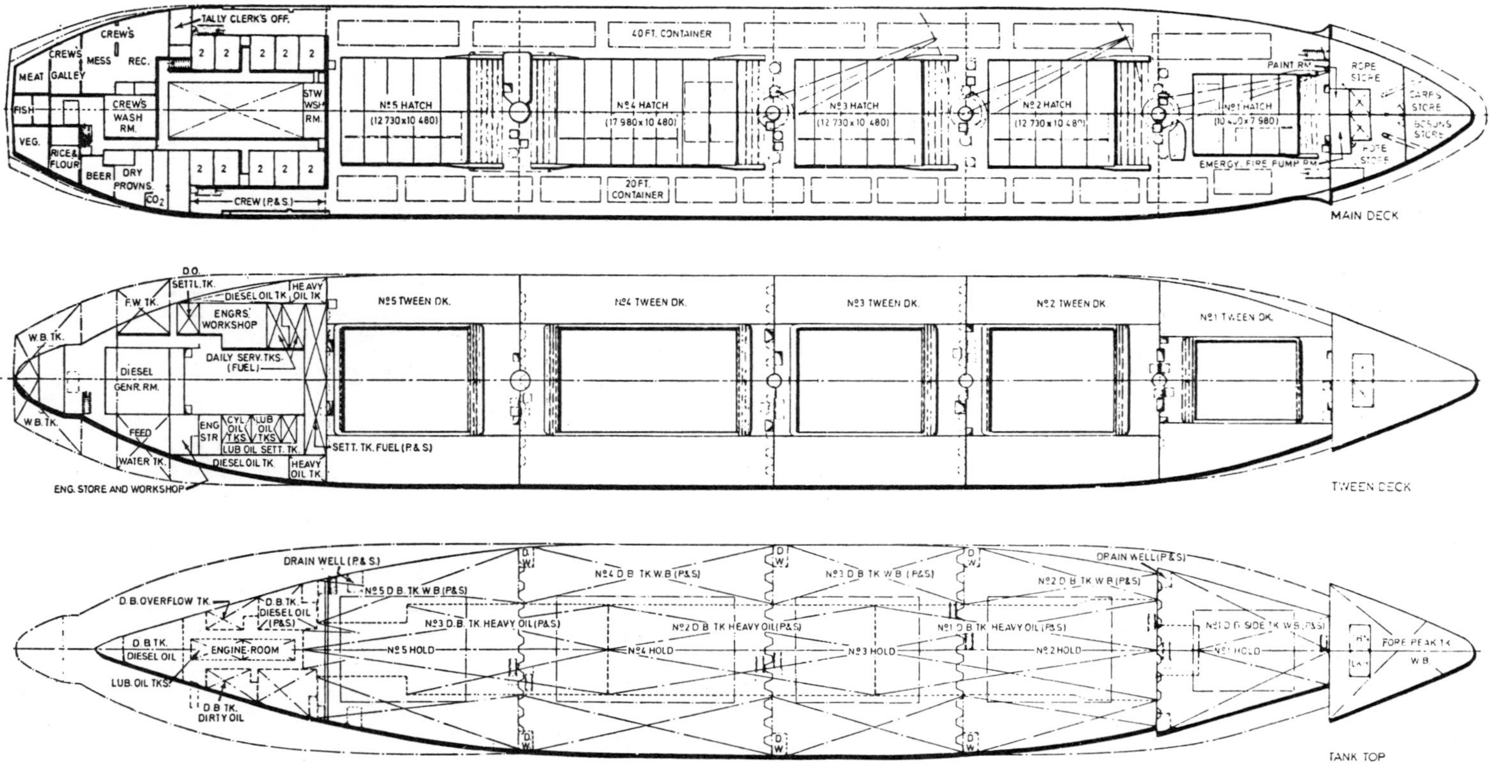


General arrangement plans of the 16,330 tonne D.W. 'Bremen Progress' class multi-purpose cargo vessels 'City of Winchester', 'City of York', 'City of Canterbury'

A traditional general cargo ship has her engine room and bridge superstructure amidships. She may have three holds forward of the bridge and two holds aft of the bridge. Forward of No. 1 hold is the forecastle and right forward is the jackstaff. Derricks are supported by masts and samson posts. They are stowed fore and aft when the ship is at sea. There are two lifeboats, one on the port side amidships, another on the starboard side amidships, abaft the funnel. The poop is situated aft and there is an ensign staff right aft.

*Exercise* 2. *Now write a description of the modern general cargo ship shown below (keep your description as close to the above description as possible):*

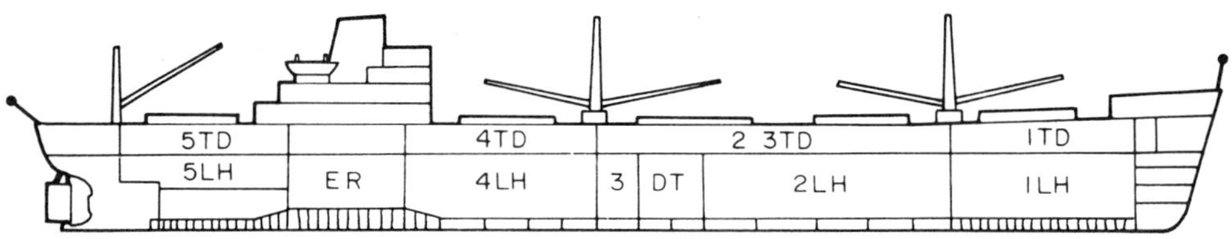


## (B) Measurement

### *(iv) Time at sea*

Time at sea is expressed in the 24-hour notation:

therefore

| | | | | |
|---|---|---|---|---|
| **on land** | 7.00 a.m. | 6.25 p.m. | midnight | 2.30 p.m. |
| **at sea** | 0700 | 1825 | 0000 | 1430 |

*Exercise* 1. *How are these times referred to at sea?*

(a) 9.00 a.m. (b) 1.30 p.m. (c) 6.25 p.m. (d) 5.45 p.m. (e) midday

*How are these times referred to on land?*
(f) 0310 (g) 1530 (h) 1115 (i) 2045 (j) 2355

Traditionally time at sea is divided into 4-hour periods called watches. There are six watches.

| | | | |
|---|---|---|---|
| Midnight–0400 | Middle Watch | Noon–1600 | Afternoon Watch |
| 0400–0800 | Morning Watch | 1600–2000 | Evening Watch |
| 0800–noon | Forenoon Watch | 2000–midnight | First Watch |

The Evening Watch can be divided into two short watches.

| | | | |
|---|---|---|---|
| 1600–1800 | First Dog Watch | 1800–2000 | Second Dog Watch |

Each watch is in the charge of an officer. The traditional pattern is as follows:

| | 12–4 | 4–8 | 8–12 |
|---|---|---|---|
| a.m. | Middle Watch | Morning Watch | Forenoon Watch |
| p.m. | Afternoon Watch | Evening Watch | First Watch |
| Deck | Second Officer | Chief Officer | Third Officer |
| Engine | Third Engineer | Second Engineer | Fourth Engineer |

Traditionally the passage of time at sea is marked by bells. Bells are struck throughout each watch as follows:

| | | | | | |
|---|---|---|---|---|---|
| After ½ hour | 1 bell | After 2 hours | 4 bells | After 3½ hours | 7 bells |
| After 1 hour | 2 bells | After 2½ hours | 5 bells | After 4 hours | 8 bells |
| After 1½ hours | 3 bells | After 3 hours | 6 bells | | |

During the Second Dog Watch the bells strike 1 2 3 8.
1 bell is struck 15 minutes before each watch is due to change.

*Exercise* 2. *Revise the above information and then answer these questions:*

(a) When is the First Watch?
(b) When is the Morning Watch?
(c) When is the First Dog Watch?
(d) What is the name of the watch between midnight and 0400?
(e) What is the name of the watch between 0800 and noon?
(f) What is the name of the watch between 1800 and 2000?
(g) Who is on duty on the bridge during the Morning Watch?
(h) Who is on duty on the bridge during the Middle Watch?
(i) Who is on duty in the engine room during the First Watch?

(j) Who is on duty in the engine room during the Afternoon Watch?
(k) What time does 4 bells in the Morning Watch indicate?
(l) What time does 1 bell in the Forenoon Watch indicate?
(m) What time does 6 bells in the First Watch indicate?
(n) What time does 3 bells in the Second Dog Watch indicate?

# GUIDED WRITING

## (A) Description of the engine department

*Write a description of the Engine Department on a cargo ship.*

*Stage* 1. *Sort out this list of engine room personnel into order of seniority:* greasers, Electrical Officer, Storekeeper, Third Engineer, Pumpman (on tankers), Chief Engineer, Donkeyman, Fourth Engineer, Engineer cadets, firemen, Second Engineer

*Stage* 2. A composition usually has an opening paragraph to introduce the subject and perhaps give an overall picture. *Read through this opening paragraph to a description of the Deck Department and then write an opening paragraph to your description using this as a guide.*

The Deck Department is made up of a number of officers, petty officers and ratings. The Chief Officer is in charge of the department and he is assisted by a Second Officer, a Third Officer and usually a Fourth Officer. There are two petty officers: a Boatswain and a Carpenter. There are also a number of deck ratings including Able Seamen, Efficient Deck Hands and Ordinary Seamen. Sometimes there are Navigating Cadets too.

*Stage* 3. *Now continue your description by writing out the following notes in a paragraph to show who is responsible for what:*

Chief Engineer – Master – the Engine Department. He (to look after) the day-to-day running of the department. Second Engineer – maintenance of the engine room, deck and other machinery. Engine room watchkeeping duties – Second, Third and Fourth Engineers. Second Engineer keeps .......... (refer to page 57 and continue in your own words). The maintenance and repair of all electrical equipment – Electrical Officer. Storekeeper and Donkeyman – Chief Engineer. Storekeeper – the storeroom.

Donkeyman – lubrication. On tankers, Pumpman – Chief Officer – loading and unloading oil and water ballast. Of the engine room ratings, Greasers – general oiling and cleaning duties, and Firemen – looking after the boilers.

You should now have a description of the Engine Department, consisting of an introductory paragraph and a main paragraph.

# NOTE-TAKING PRACTICE

## Substitutes, compounds

Here are two more ways of cutting down the amount of writing needed in note-taking.

1. By using shorter words which have a similar meaning to the long words and phrases used by the lecturer

   e.g. **can** = is/are able to, is/are capable of
   **daily** = every day, day-to-day, each day
   **more** = additional, a greater quantity of
   **better** = superior, of a higher standard/quality

2. By compressing groups of words into noun phrases. This is often done with groups of words containing **of** or **for**

   e.g. **ship's crew** = the crew of a ship
   **planning committee** = the committee for planning purposes
   **work programme** = the programme of work

   (*Note:* The forming of compound nouns is examined in a later unit.)

*Exercise* 1. *Suggest shorter words for these words and phrases:*

(a) every week
(b) accompanied by
(c) to lubricate
(d) disappeared, vanished
(e) in order to
(f) will possibly

*Exercise* 2. *Compress these phrases:*

(a) ratings of the Deck Department
(b) an appliance for extinguishing fires
(c) trials at sea
(d) the construction of ships
(e) a larger number of ratings with certificates
(f) it reduces the size of the crew

# **Lecture:** General purpose manning

A lecturer will often introduce his lecture by first telling you what his lecture is about and then telling you how he is going to deal with his subject.

*Listen to the introduction of the lecture and fill in the outline below (write down a suitable title and then note down the three parts of the lecture):*

**Title** ..............................................................................................

1. ..............................................................................................

2. ..............................................................................................

3. ..............................................................................................

A lecture is often accompanied by a handout. This may be in the form of a diagram. This diagram should help comprehension. Information can also be added to the diagram.

*Look at the diagram below as you listen to Part 1 of the lecture on the General Purpose system of manning. Add any information that you think is necessary.*

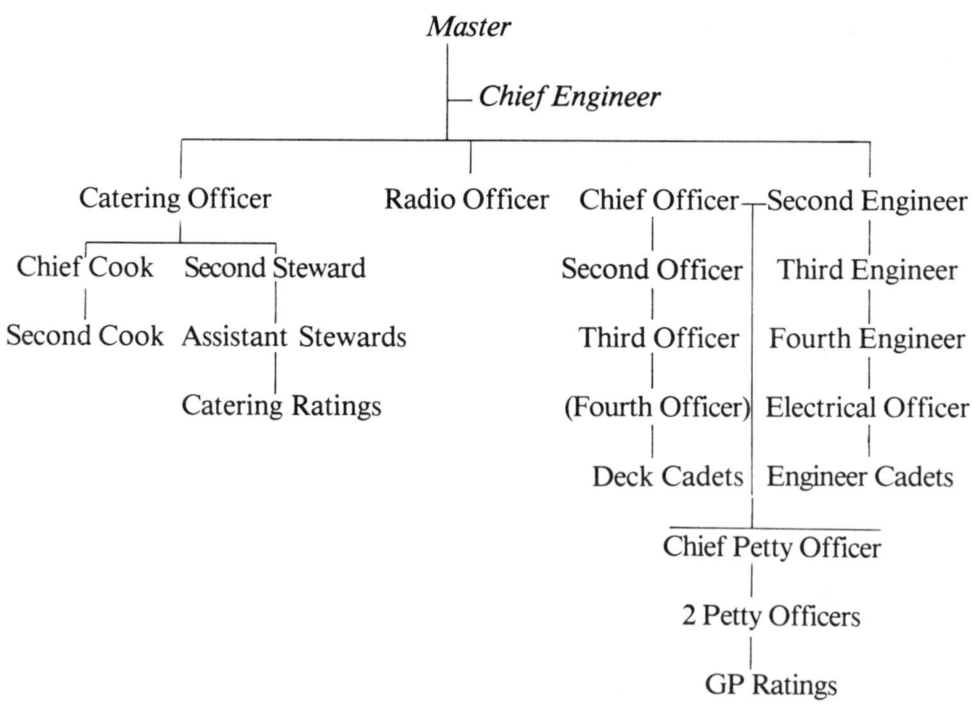

*Now listen to Part 2 of the lecture, which gives you some of the reasons for the introduction of the GP system. Take notes on the outline below:*

..............................................................................................................

1. ..............................................................................................

∴ ..............................................................................................

..............................................................................................

2. ..............................................................................................

∵ ..............................................................................................

now ..............................................................................................

∴ ..............................................................................................

..............................................................................................

..............................................................................................

*Now listen to Part 3 of the lecture, which gives you some of the advantages of the GP system of manning. Take notes on the outline below:*

..............................................................................................................

1. ..............................................................................................

(a) ..............................................................................

(b) ..............................................................................

2. ..............................................................................................

(a) ..............................................................................

(b) ..............................................................................

(c) ..............................................................................

(d) ..............................................................................

*Now turn to Guided Writing (B) and do the exercise there.*

# GUIDED WRITING

## (B) General purpose manning

*Use your notes to help you to write a description of GP manning.*

*In your first paragraph give a description of how a GP crew is made up and some of the differences between it and the traditional organization. Also say who makes up the planning committee and what it does.*

*In your second paragraph use your notes to help you give some of the reasons for the introduction of the GP system.*

*In your third paragraph use your notes to help you give some of the advantages of the system.*

*When you have finished your description, turn to the tapescript of the lecture in Appendix II. Read through the lecture and check your notes. Were your notes good enough to help you with your description?*

# UNIT V

# SEAMANSHIP

## READING COMPREHENSION

### (A) Ship handling

Diagrams and pictures are often used in technical books to show the meaning of specialized terms. They are also used to make descriptions and instructions clearer. Technical books generally try to be brief and they use diagrams to give much of the information.

*Read through the first paragraph only of the passage below and fill in the outline. This will give you the overall structure of the passage and what each paragraph is going to be about. Next read through the whole passage using the diagrams to help you understand it.*

**Title:** ..................................................................

..................................................................

1. ....................................... (a) .................................... } **under control**

(b) ....................................

2. .......................................

3. ....................................... } .........

4. ....................................... } .........

There are a number of effects to be considered when handling a ship at sea or in narrow waters. These include the axial thrust and the transverse thrust of the propeller, the effect of the rudder on the water and the effects of the wind and tide.

The effects of the propeller and rudder can be controlled. The effects of the wind and tide cannot, but their forces can be used.

The axial thrust of the propeller is the force working in a fore and aft direction. This force causes the ship to move ahead through the water or to go astern. Because of her shape, a ship will move ahead through the water more easily than going astern. See Figure 1.

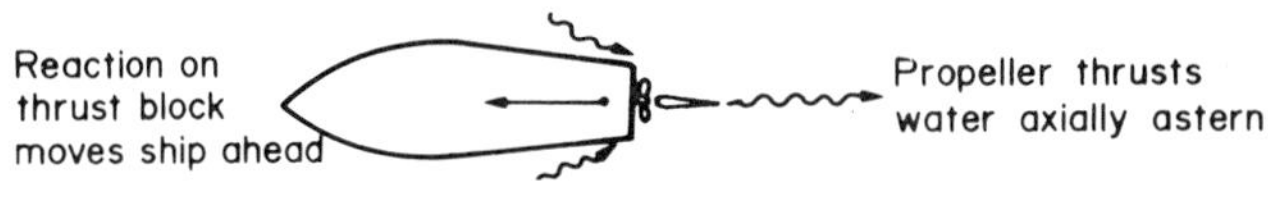


Figure 1. Axial thrust.

The transverse thrust is the sideways force of the propeller as it rotates. The transverse effect of the propeller blades at the top near the surface of the water is not strong enough to counteract the opposite effect of the lower blades. For right-handed propellers this cants the ship's stern to starboard and her bow to port, when the ship is going ahead. The effect is small and can be corrected by the rudder. When the engines are put astern, the effect is the opposite and the stern cants to port. This effect is stronger and cannot easily be corrected. Vessels with left-handed propellers behave in the opposite way to that shown in Figure 2.

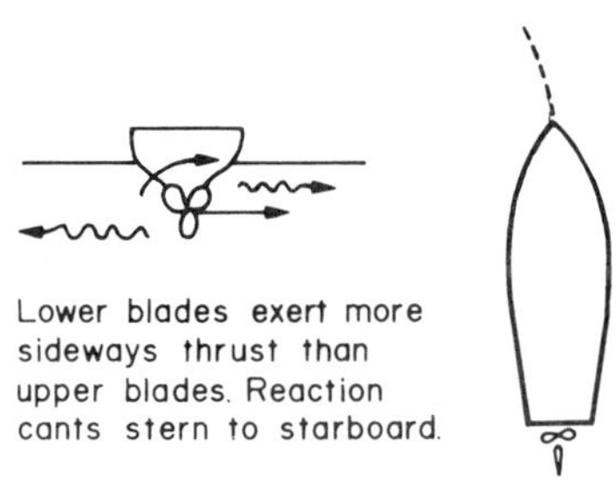


Figure 2. Transverse thrust ahead.

The rudder depends for its effect on the deflection of a stream of water. The stream is produced by the ship's movement through the water and by the axial flow of water from the propeller. When the rudder is put to starboard, the stream of water is deflected to starboard. This cants the stern to port and the ship's head turns to starboard (see Figure 3). When the rudder is put to port the effect is the opposite. When the engine is put astern the effect of the rudder is unpredictable.

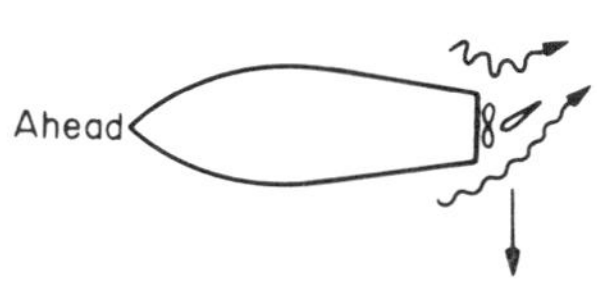


Figure 3. Rudder action.

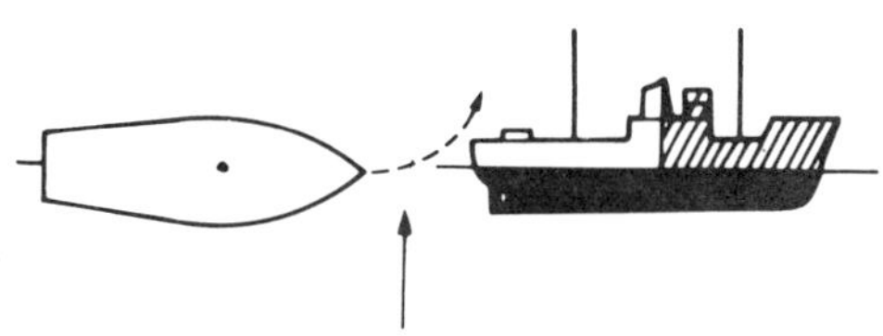

Figure 4. More windage for'd.

The effect of the wind hitting a ship sideways on depends on the area presented to the wind before or aft of her pivoting point. Ships with their main superstructure amidships will turn away from the wind (see Figure 4). Ships with the main superstructure aft, such as tankers, will turn towards the wind.

A beam wind will also cause a ship to make leeway. That is, she will be blown sideways off her course away from the wind as she moves forward through the water (see Figure 5).

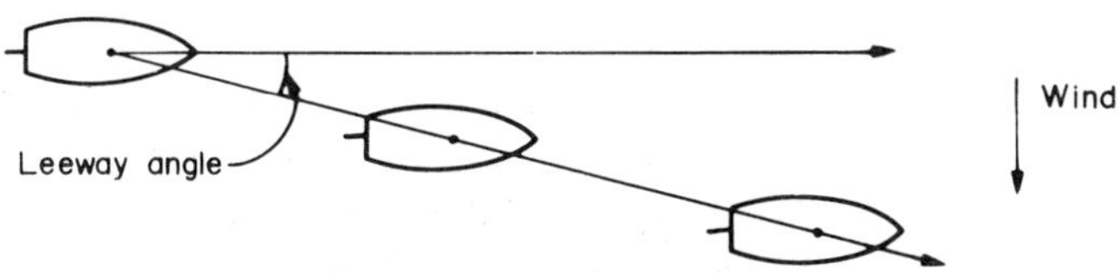


Figure 5. Beam wind action.

Current and tide also affect a ship. The current will carry the ship with it, or slow it down when the ship is travelling against the current. The tide will lift the ship and lower it. It will also carry it in its direction and cause the ship to turn when anchored or moored.

*Using the diagrams of rudder action and wind effect as guides, draw diagrams to illustrate:* (1) *the deflection of water and movement of the ship when the rudder is put to port;* (2) *the effects of a beam wind on a tanker.*

# GRAMMAR

## (A) Articles

### *(ii) The, zero article with geographical names*

*Study the lists below of* *<u>when</u>* *and* *<u>when not</u>* *to use the definite article* ***(the)*** *with geographical names and terms:*

1. Generally **the** is *not* used with names of

| | |
|---|---|
| **Continents** | Asia, Europe, South America (**except**: The Arctic, The Antarctic) |
| **Countries** | Greece, Kuwait (**exceptions include:** The Netherlands, The Lebanon) |
| **Islands** | When thought of as one island: Japan, Cuba, Hawaii |
| **Cities** | Alexandria, Ankara, New York, London |
| **Towns** | Dover, Brighton, Calais (**except:** The Hague) |
| **Mountains** | Mount Everest, Mont Blanc, Ben Nevis (**exceptions include:** The Jungfrau, The Eiger) |

**Lakes** Lake Michigan, Lake Como, Loch Ness
**Bays** Raleigh Bay, Galway Bay (**except when:** The Bay of . . .)

(*Note:* Although there are exceptions, **the** is not used with the above except when referred to in the following way: The City of London, The Bay of Biscay. Countries with **the** were once thought of as areas (see below) rather than countries. They often have two names e.g. The Netherlands – Holland, The Argentine – Argentina.)

2. Generally **the** is used with the names of

| | |
|---|---|
| **Countries** | *when abbreviated:* the USA, the USSR, the UAE, the UK |
| **Areas** | *of countries:* the North East; *of cities:* the West End |
| **Deserts** | *when large:* the Sahara, the Kalahari |
| **Groups of islands** | The Philippines, The Canaries |
| **Ranges of mountains** | The Alps, the Himalayas, the Rockies |
| **Oceans/seas** | The Pacific, the Atlantic, the Mediterranean Sea |
| **Channels/straits** | The English Channel, the Straits of Gibraltar |
| **Gulfs** | The Persian Gulf/the Arabian Gulf, the Gulf of Oman |
| **Estuaries** | The Amazon Estuary, the Nile Delta |
| **Currents** | The Agulhas Current, the North Atlantic Drift |
| **Rivers/canals** | The Nile, the Euphrates, the Suez Canal |

*Exercise* 1. *Look through a good world atlas and find other examples of the names of continents, countries, etc. Note if the article* (**the**) *is used or not. Can you find any exceptions to the above patterns of use?*

*Exercise* 2. *Write a short description of your own country based as far as possible on the description below:*

Scotland is part of Britain. Along with England, Wales and Northern Ireland, it makes up the UK. There are many off-shore islands. These include Skye and Arran in the west and the Orkneys in the north. The largest range of mountains is the Grampians and the highest mountain is Ben Nevis. The capital is Edinburgh, but Glasgow has the largest population.

## (B) Time relaters

### *(iii) After, when, while, on, during, until*

Here are some more ways of introducing time clauses.

1. When one event immediately follows another **when** can be used instead of **after**

   e.g. After a ship arrives in port, port officials go aboard.
   ⇒ When a ship arrives in port, port officials go aboard.

2. When the subject is the same in both clauses **when** can often be rewritten with **on + noun/-ing**

   e.g. When the customs men arrive on board, they go to the Captain's cabin.
   ⇒ On arriving on board, the customs men go to the Captain's cabin.

   e.g. When the cargo was unloaded, it was put into the warehouse.
   ⇒ On being unloaded, the cargo was put into the warehouse.

3. When one event happens at the same time as another, **when/while** can sometimes be replaced by **during + noun**

   e.g. While a ship is being launched, tugs stand by ready to help.
   ⇒ During the launching of a ship, tugs stand by ready to help.

4. When we want to show the limit of an action, we use **until**

   e.g. The Chief Officer stayed on watch until 0800.

   If we are referring to the time the action began, we must put the verb in the main clause in the *negative.*

   e.g. The Chief Officer did not go on watch until 0400.

*Exercise* 1. *Rewrite the following sentences using the words in brackets.*

(a) After a ship is completed, she undergoes speed trials. **(When)**
(b) While a ship is being constructed, she is continually being examined. **(During)**
(c) He went to sea when he was eighteen years old. **(Until)**
(d) After the passengers arrive on board, they are shown to their cabins. **(On)**
(e) While a cadet is being trained, he is paid by his company. **(During)**
(f) The passengers went below when the sun went in. **(Until)**
(g) When the hull was inspected, it was found to be damaged. **(On)**
(h) When a vessel undergoes speed trials, her maximum speed is measured. **(During)**
(i) The pilot disembarked when the ship was clear of the harbour. **(Until)**

# READING COMPREHENSION

## (B) Ship handling (cont.)

Here are some more passages showing how texts and diagrams work together to convey meaning.

*Read this passage and do the exercises as you come to them:*

A ship is made fast to the quayside by mooring lines. The standard mooring lines are shown below. They consist of a headline, a breastline and a backspring forward, a stern line, a breastline and a backspring aft. Any of these lines may be doubled. Each line has a large eye spliced in the end. The eye is placed over a bollard on the quayside. If there is another line already on the bollard, the eye of the second line should be taken up through the eye of the first line before placing it over the bollard. This makes it possible for either line to be let go first.

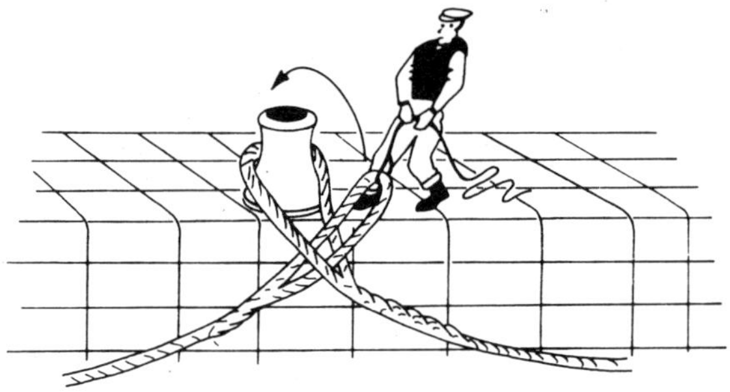

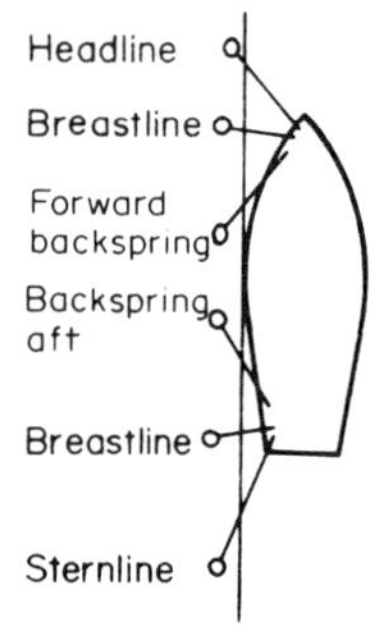


Putting eye on bollard

**Leaving berth port side to** (*no wind or tide*). Single up to a backspring forward and put engines to slow ahead. Put rudder hard to port (1). This cants the stern out away from the berth (2). When about 30° out, stop engines and put rudder amidships and engines half astern. Let go the backspring as the vessel moves astern off the berth (3). The effect of the transverse thrust is to take the stern to port and the bows will swing to starboard clear of the berth. Stop engines and then go full ahead.

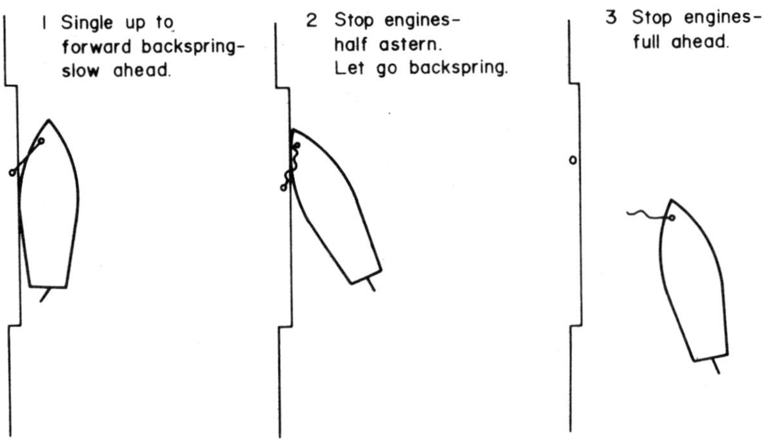


Leaving berth (port side to)

*Now use the diagram and notes below to help you complete this description of a ship leaving a berth, starboard side to, with no wind or tide.*

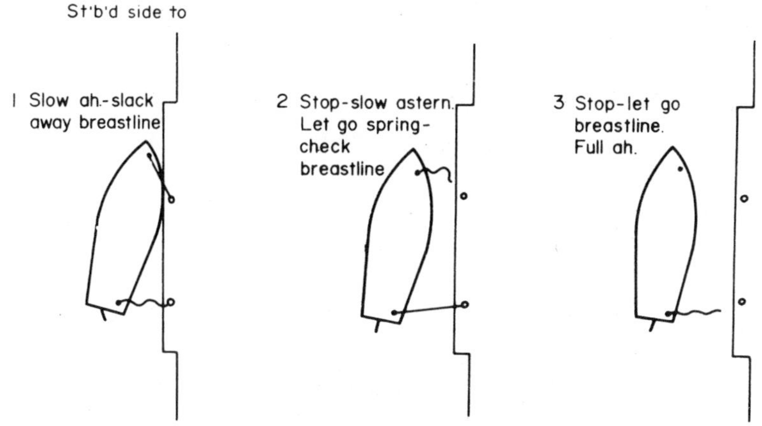


Leaving berth (starboard side to)

**Leaving berth starboard side to** (*no wind or tide*). Single up to a ............... forward and a ............... aft. Put the engines to ............... and the rudder hard to ............... (1). Slack away on the ............... until the stern is clear of the berth (2). Stop engines, then go ..............., put the rudder ............... . Let go the ............... and check on the ............... . The action of the ............... is to prevent transverse thrust taking the stern to ............... and consequently forcing the bows onto the quayside. When all is clear ............... engines and let go the ...............(3). Then put engines ............... .

*Complete the description using only the diagram to help you.*

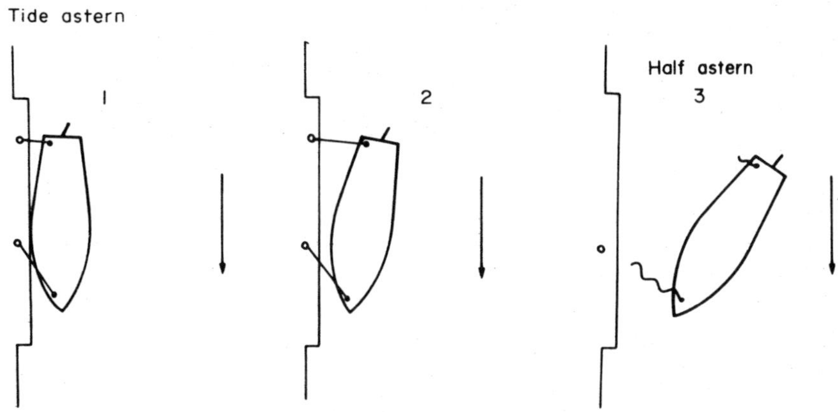


Leaving berth tide astern

**Leaving berth tide astern.** Single up to a .............. forward and a .............. aft(1). Slack away on the .............. and the tide will force the vessel away from the berth. This can be helped by putting the rudder hard over away from the berth, in this case, hard to ................ (2). When clear of the berth, let go the ................ and the .............., and put the engines .............. and the rudder .............. .

# APPLIED TERMINOLOGY

## (A) Terms relating to a ship's movement

1. The motions of a ship

*Study the diagram below showing the six motions of a ship:*

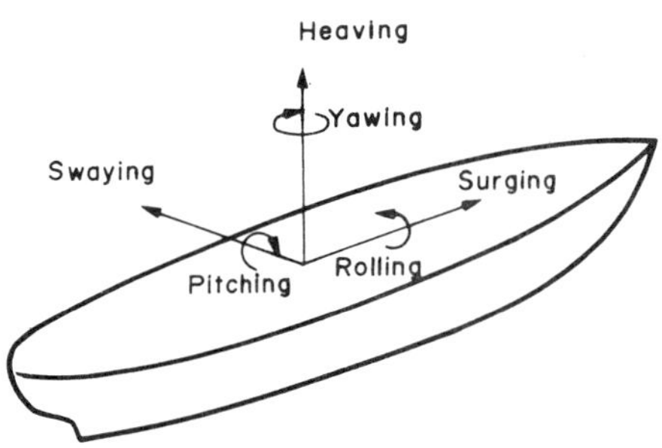


Ship in motion

*Exercise* 1. *Using the diagram to help you, complete these definitions:*

(a) A ship is said to be .............. when the bow and stern rise and fall with the oncoming waves.

(b) A ship is said to be .............. when the port and starboard sides rise and fall with waves coming from abeam.

(c) A ship is said to be .............. when the whole ship rises bodily and then sinks into the trough of a wave.

(d) A ship is said to be .............. when she is pushed forward by the waves and seems to move backwards as she falls back into the troughs.

(e) A ship is said to be .............. when the whole ship moves bodily to port and then to starboard.

(f) A ship is said to be .............. when her bow is pushed first to port and then to starboard.

2. The movement of a ship through the water

*Study this definition:*

A ship is said to be afloat when she is borne by the water.

*Exercise 2.* *Now try to write definitions of the verbs below using the same pattern. The first has been done for you.*

| Verb | Meaning |
|---|---|
| (a) to be afloat | borne by the water |
| (b) to be underway | neither anchored, moored, made fast, nor aground. |
| (c) to be making headway | moving forwards through the water |
| (d) to be making sternway | moving backwards through the water |
| (e) to be making leeway | moving sideways through the water being blown by the wind |
| (f) to overhaul (another vessel) | overtaking her |
| (g) to fall astern | dropping behind a faster vessel |
| (h) to be hove to | stopped at sea |
| (i) to be adrift | moving without means of propulsion |
| (j) to be aground | lying on the bottom or stuck on rocks |
| (k) to bear away | turned away from the wind |
| (l) to be moving broadside on | moving sideways through the water |

## (B) Measurement

### *(v) Depth*

The traditional method of measuring depth at sea is to use the **Hand leadline** (almost never used on modern ships). It consists of a length of line which is marked in fathoms (1 fathom = 6 feet) or metres as follows:

| | |
|---|---|
| 2 fathoms – two leather strips | 2 metres – two leather strips |
| 3 fathoms – three leather strips | 3 metres – blue bunting |
| 5 fathoms – white linen | 5 metres – white bunting |
| 7 fathoms – red bunting | 7 metres – red bunting |
| 10 fathoms – piece of leather with a hole in it | 10 metres – a piece of leather with a hole in it |
| 13 fathoms – blue serge | 13 metres – blue bunting |
| 15 fathoms – white linen | 15 metres – white bunting |
| 17 fathoms – red bunting | 17 metres – red bunting |
| 20 fathoms – length of cord with two knots | 20 metres – piece of leather with a hole in it, and two strips of leather |

When calling the lead, the marked fathoms (see above) are called *marks*, while the unmarked fathoms (1, 4, 6, 8, 9, 11, 12, 14, 16, 18, 19) are called *deeps*. Soundings are called to quarter fathoms but the words 'three-quarters' are never used.

The calls are given as follows:

e.g. At nine fathoms – 'By the deep nine'
At nine and a quarter fathoms – 'And a quarter nine'
At nine and a half fathoms – 'And a half nine'
At nine and three quarter fathoms – 'A quarter less ten'
At ten fathoms – 'By the mark ten'

If no bottom is reached when the line is up and down, the call should be 'No bottom at *x* fathoms'.

*Exercise 1. What would the calls be for the following depths?*

(a) 5 fathoms (b) 8½ fathoms (c) 16 fathoms (d) 7¼ fathoms
(e) 13¾ fathoms (f) 2 fathoms (g) 9¼ fathoms (h) 3¾ fathoms

# GUIDED WRITING

## (A) Description of weighing anchor

*Write a description of what to do when weighing anchor.*

*Stage* 1. *Read this description of what to do when preparing to anchor. Use the diagram to help you with vocabulary.*

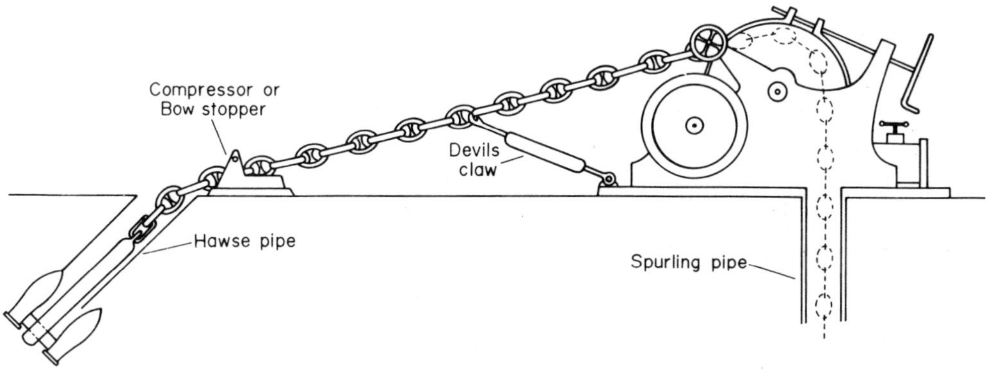


When preparing to anchor, it is necessary to take to the forecastle a hammer, an oil can and some goggles. Before going to the forecastle, ask the engine room for power and water on deck. First, take off the covers from the hawse pipes and clear the spurling pipes. Then, make sure that the windlass is out of gear and the brakes are on. Next, turn the windlass over slowly. While you are doing this, oil all the moving parts. On a steam windlass, the drain cocks must be opened and the water drained off before the cocks are closed again. After seeing that the gears are free to engage, put the windlass into gear. Then remove the devil's claw and the compressor bar. On receiving the order to lower away, lower the anchor until it is clear of the hawse pipe. When you are sure that the weight of the anchor is held by the brake, take the windlass out of gear. When the order to let go the anchor is received, release the brake. As each shackle (27.5 metres) goes out, the forecastle bell is rung (1 ring for 1 shackle, 2 rings for 2 shackles etc.). When the anchor reaches the bottom, hoist the anchor ball or anchor lights. When enough cable has been paid out, apply the brake and put on the compressor bar. Then inform the engine room that power and water are finished with. Finally return all the gear.

*Stage 2.* *Now read through this list of instructions and events for weighing anchor. Make sure you understand each stage.*

1. Take to the forecastle a hammer, an oil can and a hose.
2. Ask the engine room for power and water on deck.
3. Couple up the hose, lead it to the hawse pipe and open the water cock.
4. Make sure that the windlass is out of gear and the brakes are on.
5. Turn the windlass over slowly.
6. Oil the moving parts.
7. See that the gears are free to engage.
8. Put the windlass into gear.
9. Remove the compressor bar.

10. The order to weigh anchor is received.
11. Release the brake and begin heaving in the cable.
12. The cable is heaved aboard.
13. The cable is hosed down to wash away any dirt.
14. The shackles appear.
15. The right number of rings is given on the forecastle bell.
16. Ring the forecastle bell rapidly and haul down the anchor ball or anchor lights.
17. This is done when the anchor is aweigh.
18. The anchor is hove home.
19. Apply the brake and replace the compressor bar.
20. Take the windlass out of gear.
21. Inform the engine room that power and water are finished with.
22. Return all the gear.

*Stage* 3. *Using the first description as a guide, join together the above list of sentences to form a description of what to do when weighing anchor. Keep your description as close as possible to the first, using the same words for expressing time relations.*

# NOTE-TAKING PRACTICE

## New vocabulary

When we take notes we have to do three things:
1. We have to understand what the lecturer is saying.
2. We have to decide what the important information is.
3. We have to write this down briefly and coherently.

So far we have looked at a number of techniques for reducing the amount of writing (e.g. abbreviations, symbols, omissions etc.) and a number of ways of organizing notes on paper (e.g. numbering paragraphs: I, (i), 1, (a) etc.).

Understanding a talk or lecture can be very difficult because you have to concentrate for long periods and there may be many words you do not recognize. One way of helping with the problem of new vocabulary is to prepare for a lecture by reading a little about the subject beforehand. For example, you can make a list of words relevant to the subject and find out their meaning. There is a good chance that the lecturer will use some of these.

*The following words are likely to be used in a lecture on different types of rope. Use a good dictionary to find out their meaning:*

fibre; natural; synthetic; wire; flexible; lashings; to shrink; to swell; buoyant; elastic; to rot; to tow; yachts; water-resistant; to melt; log lines; halyards; galvanized; to rust.

## **Lecture:** Different types of rope

*Listen to the introduction to the lecture and fill in the outline below. Write down a title and then suitable headings to the three main parts of the lecture.*

**Title:** ...........................................................................................

1. ...........................................................................................

2. ...........................................................................................

3. ...........................................................................................

*Now listen to the rest of the lecture. It should be played three times. The first time, concentrate on listening for the basic type of rope and the individual names. Note these down on the outline in the space provided. The second time, note down the characteristics of each type of rope and the third time, some of the uses.*

| **Basic types** | **Names** | **Characteristics** | **Uses** |
|---|---|---|---|
| 1. ............ | (a) ............ | .............................. | .............................. |
| | | .............................. | .............................. |
| | (b) ............ | .............................. | .............................. |
| | | .............................. | .............................. |
| | (c) ............ | .............................. | .............................. |
| | | .............................. | .............................. |
| | (d) ............ | .............................. | .............................. |
| | | .............................. | .............................. |

(e) ............ ............................... ...............................

............................... ...............................

2. ............ (a) ............ ............................... ...............................

............................... ...............................

(b) ............ ............................... ...............................

............................... ...............................

(c) ............ ............................... ...............................

............................... ...............................

3. ............ (a) ............ ............................... ...............................

## GUIDED WRITING

### (B) Different types of rope

*Use the notes you have taken to write an essay on the different types of rope, their characteristics and their uses.*

*Your essay should have an introduction and three paragraphs, one paragraph on each basic type of rope.*

*When you have finished, turn to Appendix II at the back of the book and compare your essay to the tapescript.*

# UNIT VI

# CARGO WORK

## READING COMPREHENSION

### (A) Different types of cargo (version 1)

*Read through the following passage on the different types of cargo. Use the techniques that you have practised so far to help you.*

Merchant ships are designed to carry cargo. This cargo may be divided into two basic types: bulk cargo and general cargo. Bulk cargo consists of a single commodity. This commodity is usually carried loose. General cargo consists of a variety of goods. These goods are packed separately. Bulk cargo is carried in specially designed vessels, therefore stowage presents few problems. With general cargo stowage presents many problems, because each item has its own type of packaging and characteristics.

Bulk cargo can be divided into liquid or dry bulk cargo. Liquid bulk cargo is carried in tankers. Most tankers are designed to carry crude oil or its refined products, such as fuel oils. The oil is carried in tanks. These are connected by a system of pipes to a central manifold. The cargo is pumped on board at the loading port by shore pumps. At the discharging port the ship pumps the oil ashore using her own pumps. Dry bulk cargo is carried in bulk carriers. The cargo is carried in self-trimming holds. Dry bulk cargo includes grain, iron-ore, coal and sugar. It is loaded automatically by buckets on a conveyor belt system or through large tubes. Although the cargo stows itself, it is important to maintain the ship's stability and to make sure that the cargo will not move during the voyage. Dry bulk cargo is unloaded by huge grabs on cranes or by giant suction tubes.

General cargo can be divided into containerized, non-containerized and refrigerated cargo. Non-containerized cargo presents most stowage problems, because each commodity has its own type of packaging and characteristics. Goods may be in bags, bales, cases or steel drums. Individual pieces of machinery may not be packaged at all. Some cargoes such as tobacco and rubber have a strong odour and will taint delicate cargoes such as tea and rice. Other cargoes such as cement and fertilizers are dusty and leave a residue behind them. Heavy cargoes must not be

A general cargo carrier using her own derricks (*Unit VI*)

Loading a bus (*Unit VI*)

Unloading coal with a grab (*Unit VI*)

Containers being loaded (*Unit VI*)

A 53,000 ton tanker, discharging at Angle Bay Ocean Terminal, Pembrokeshire (*Unit VI*)

A dry bulk carrier unloading (*Unit VI*)

stowed on top of fragile cargoes. This can cause problems if the heavy cargo has to come out first. General cargo is loaded by cranes and the ship's own derricks. Non-containerized cargo is carried in multi-deck vessels. To help with the problem of stowage many types of general cargo are now being put into containers of standard dimensions. A container is 8 feet high and 8 feet wide (2.44m × 2.44m) and is usually 20 feet or 40 feet (6.1m or 12.2m) in length. They are carried in specially designed container ships and loaded and unloaded by special cranes from the quayside. The containers are stowed both above and below deck. Perishable cargoes such as meat, fruit and dairy produce are carried in ships with refrigerated holds. These holds are designed to keep food at the correct temperature. Some food such as fish is frozen solid, other food such as fruit is only chilled. Mutton and lamb are stowed fore and aft, beef when chilled is hung on hooks. Eggs and butter are easily tainted. Fruit needs good ventilation. Refrigerated cargo is loaded by cranes and derricks.

*Complete this diagram to form a summary of the basic types of cargo:*

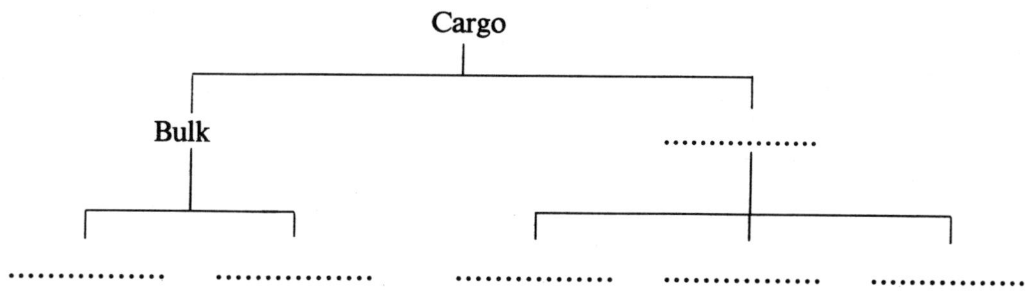


*Now complete this table to summarize some more information in the passage:*

| Type of cargo | Examples | Type of ship | How (un)loaded |
|---|---|---|---|
| liquid bulk | ............................ | ............................ | ............................ |
| dry bulk | ............................ | ............................ | ............................ |
| containerized | ............................ | ............................ | ............................ |
| non-containerized | ............................ | ............................ | ............................ |
| refrigerated | ............................ | ............................ | ............................ |

# GRAMMAR

## (A) Prepositions of place (i)

*A preposition can have several meanings. Study how the following prepositions are used to describe position:*

1. **At/away from** – used with reference to a point

| ● **at** | ● **away from** |
|---|---|
| The helmsman stood at the wheel. | The engineer stood away from the flames. |
| The ship refuelled at Dakar. | The ship anchored away from other ships. |

2. **On/off** – used with reference to a line or surface

| **on** | **off** |
|---|---|
| The ship stayed on course. | The ship was off course. |
| The chart lay on the table. | The keel was just off the seabed. |

3. **In/out of** – used with reference to an area or volume

| **in** | **out of** |
|---|---|
| The ship moored in the harbour. | The instruments lay out of their box. |
| Spirits are kept in lock-ups. | He stood out of the way of the derrick. |

*Exercise* 1. *Complete these sentences with a preposition from above:*

(a) Meat is carried ........... refrigerated holds.

(b) The photograph hung ........... the cabin wall.

(c) The ship sank a mile ........... shore.

(d) The cargo was stowed ........... the boilers.

(e) The tanker took on supplies ........... Cape Town.

(f) The sailor sat ........... the heat of the sun.

(g) Ammunition is kept ........... magazines.

(h) The ship's company formed up ........... the quarter deck.

*Now study how these prepositions are used to describe relative position:*

4. **Above/below** means on a higher or lower level

| | |
|---|---|
| **above** | The clouds are above the horizon. |
| **below** | Cargo is usually stowed below deck. |

5. **Over/under** – implies a direct vertical relationship

| | |
|---|---|
| **over** ↕ | The helicopter hovered over the platform. |
| **under** | The student put the flame under the flask. |

6. **On top of/underneath** suggests contact or concealment

| | |
|---|---|
| **on top of** | His suitcase lay on top of his bunk. |
| **underneath** | The spanner was found underneath the tarpaulin. |

7. **Behind/in front of** (compare *abaft/aft* of and *before/forward* of)

| | |
|---|---|
| **Behind \| in front of** | The navigator sat behind the pilot.<br>He stood in front of the mirror. |

*Exercise* 2. *Complete this description with a preposition from 1 to 7 as appropriate:*

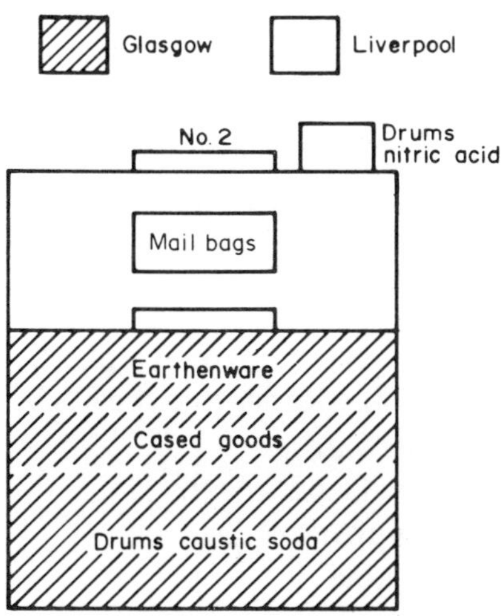


The cargo for Liverpool and Glasgow is stowed ......... No. 2 hold. The cargo for Liverpool is stowed ......... the cargo for Glasgow, because it is to be unloaded first. Six drums of nitric acid to be unloaded ......... Liverpool are stowed ......... deck. Mail bags are stowed ........ the hatch square of the 'tween deck. ......... the lower hold cased goods are stowed ......... drums of caustic soda. Directly ......... the cased goods is earthenware.

## (B) Relative clauses (i)

Structures can be linked by using defining and non-defining relative clauses. *Study how they are used in the examples below:*

1. Defining relative clauses

(a) Some ships are called bulk carriers.
(b) These ships carry bulk cargo.
(a) + (b) Ships which carry bulk cargo are called bulk carriers.

(a) Cotton comes from a plant.
(b) This plant grows in the USA.
(a) + (b) Cotton comes from a plant which grows in the USA.

(a) A man fell overboard.
(b) The man was painting a lifeboat.
(a) + (b) The man who was painting a lifeboat fell overboard.

Sentences (a) are general and undefined. Sentences (b) tell us which 'ships', 'plant', and 'man' is being referred to. That is why they are called defining clauses. They are put immediately after the noun they define. **Who** is used for people, **which** for things. There is no comma (,) before them.

*Exercise* 1. *Join these sentences together as shown in the examples above:*

(a) Some ships are called colliers.
These ships carry coal.
(b) Hemp and sisal rope are made from the leaves of a plant.
This plant grows in parts of Europe, Russia and the USA.
(c) One officer is called the Chief Steward.
He is responsible for the Catering Department.
(d) Some cargoes are stowed in lock-ups.
These cargoes are classed as valuable.
(e) An officer may become the Master of his own ship.
He has the right qualifications and experience.

2. Non-defining relative clauses

(a) All the deck cargo was unloaded first.
(b) The cargo consisted of steel rails.
(a) + (b) All the deck cargo, which consisted of steel rails, was unloaded first.

(a) The First Mate is responsible for the cargo.
(b) The First Mate must have at least a class 2 Certificate.
(a) + (b) The First Mate, who must have at least a class 2 Certificate, is responsible for the cargo.

Here sentences (b) do not define the 'deck cargo' or the 'First Mate', they just add extra information about them. This information can be left out without changing the meaning of the sentence. For this reason commas (,) are put round the relative clause.

*Exercise* 2. *Read through this description of the Suez Canal. Then add the information below in the appropriate place using non-defining relative clauses.*

The Suez Canal was opened on 17 November, 1869. It was designed by Ferdinand de Lesseps. The Canal joins the Mediterranean Sea to the Red Sea. Although it is twice as long as the Panama Canal, it cost half as much to build. This was because Suez is a sea-level canal from end-to-end, therefore locks are unnecessary. The Canal has been widened and deepened several times since it was opened. It is now about twice its original breadth and depth. Until 1956, the Canal was operated by the Suez Canal Company. Since this date it has been operated by the Egyptian government.

(a) The Suez Canal took ten years to build.
(b) Ferdinand de Lesseps was a French engineer.
(c) The Canal is 105 miles long.
(d) Locks are expensive to build.
(e) The Canal was originally 58 metres wide at the surface and about 6 metres deep.
(f) The Egyptian government plan to increase its depth to take super-tankers.

# READING COMPREHENSION

## (B) Different types of cargo (version 2)

*Read through the passage below, which is a development of the reading passage in Reading Comprehension (A). Here defining and non-defining relative clauses have been used to rewrite some sentences and to add extra information to others. As you read through the passage, write in the margin Df or N-Df opposite each clause as appropriate. The first two have been done for you. Also circle the words and phrases that the words underlined refer to, to show the links between and within sentences.*

Df [1]Merchant ships are designed to carry cargo. [2]The cargo which they carry may be
divided into two basic types: bulk cargo and general cargo. [3]The former consists of
N-Df a single cargo, which is usually carried loose. [4]The latter consists of a variety of goods, which are packed separately. [5]Bulk cargo is carried in specially designed vessels, therefore stowage presents few problems. [6]With general cargo stowage presents many problems, because each item, which is different in shape and size, has its own type of packaging and characteristics.

[7]Cargo which is carried in bulk can be divided into liquid or dry bulk cargo. [8]Liquid bulk cargo is carried in tankers. [9]Most are designed to carry crude oil, which is transported to the refineries, or its refined products, such as fuel oils. [10]The oil is carried in tanks which are connected by a system of pipes to a central manifold. [11]The cargo is pumped on board at the loading port by shore pumps. [12]At the discharging port the ship pumps the oil ashore using her own pumps, which may be of the reciprocating or the centrifugal type. [13]Dry bulk cargo is carried in bulk carriers. [14]The cargo, which includes grain, iron-ore, coal and sugar, is carried in self-trimming holds. [15]It is unloaded automatically by buckets on a conveyor belt system or through large tubes. [16]Although the cargo stows itself, it is important to maintain the ship's stability and to make sure that the cargo will not move during the voyage. [17]Dry bulk cargo is unloaded by huge grabs on cranes or by giant suction tubes which are called elevators.

[18]General cargo can be divided into containerized, non-containerized and refrigerated cargo. [19]Cargo which is not in containers presents the greatest stowage problem, because each commodity has its own type of packaging and characteristics. [20]Goods may be in bags, bales, cases or steel drums. [21]Individual pieces of machinery which are large and awkward may not be packaged at all. [22]Cargoes which have a strong odour, such as tobacco and rubber, will taint delicate cargoes, such as tea and rice. [23]Cargoes which are dusty, such as cement and fertilizers, leave a residue behind them. [24]Heavy cargoes must not be stowed on top of fragile ones. [25]This can cause problems if the heavy cargo has to come out first. [26]General cargo which is not in containers is carried in multi-deck vessels. [27]To help with the problem of stowage many types of general cargo are now being put into containers of standard dimensions. [28]Containers, which are 8 feet high and 8 feet wide (2.44m × 2.44m) and 20 feet or 40 feet (6.1m or 12.2m) in length, are carried in specially designed container ships and loaded and unloaded by special cranes from the quayside. [29]The containers are stowed both above and below deck. [30]Cargo which is perishable, such as meat, fruit and dairy produce, is carried in ships which have refrigerated holds. [31]These are designed to keep food at the correct temperature. [32]Some food, such as fish, is frozen solid, other food, such as fruit, is only chilled. [33]Mutton and lamb are stowed fore and aft, beef when chilled is hung from hooks, which slide along rails. [34]Eggs and butter, which are packed in cases, are easily tainted. [35]Fruit needs good ventilation. [36]Refrigerated cargo is loaded in the same way as general cargo which is non-containerized.

*Now answer these questions in complete sentences:*

(a) What is bulk cargo?
(b) What is general cargo?
(c) Why is bulk cargo easier to stow than general cargo?
(d) What type of pumps may tankers have for unloading oil?
(e) What type of cargoes will taint delicate cargoes?
(f) Why are general cargoes now being put into containers?
(g) What are the dimensions of a container?
(h) What sort of cargo is carried in refrigerated holds?

# APPLIED TERMINOLOGY

## (A) Terms relating to cargo handling gear

*Exercise* 1. *Look at these pictures of cargo-handling gear and then write down what each of them is designed to lift (the first has been done for you):*

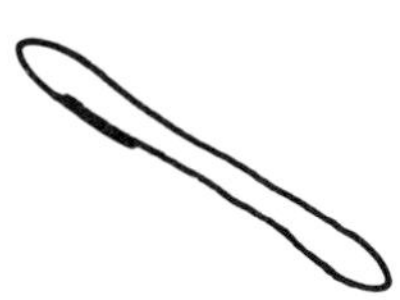

(a) Rope sling
(bags, bales)

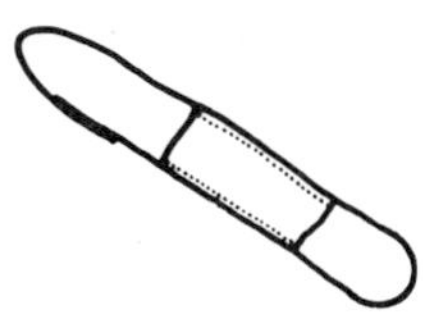

(b) Canvas sling
(bags of grain, rice, coffee)

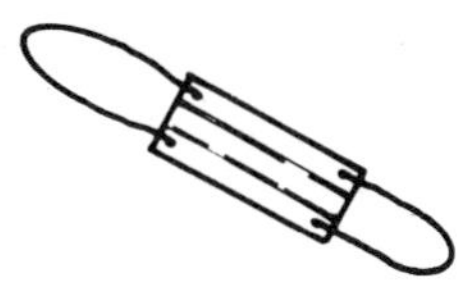

(c) Board sling
(bags of cement)

(d) Snotter
(cases, bales, hides, timber)

(e) Chain sling
(logs, iron rails)

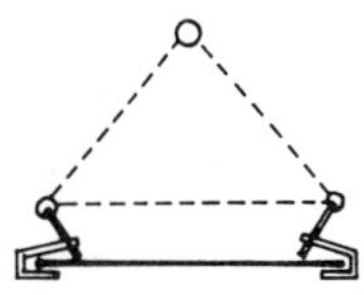

(f) Plate clamps
(steel plates)

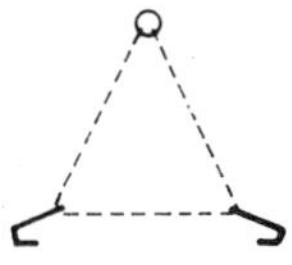

(g) Can hooks
(drums, barrels)

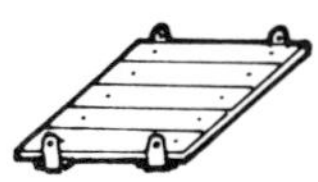

(h) Tray
(cases, drums)

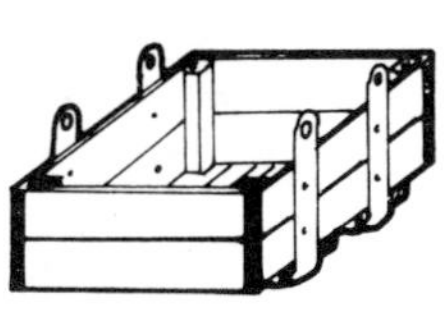

(i) Box
(explosives)

(j) **Net** (small packages, mail)

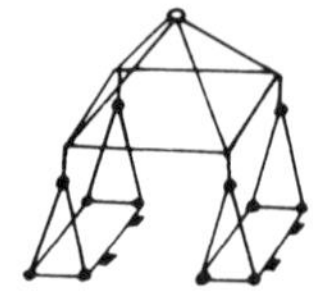

(k) **Car sling** (cars, lorries)

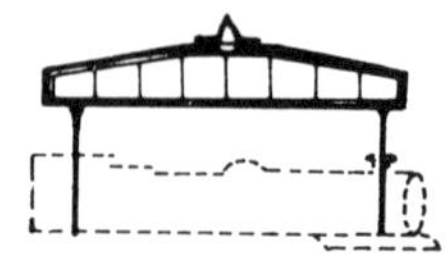

(l) **Heavy lift beam** (locomotives, long heavy cargoes)

(a) A rope sling is used for lifting bags and bales.

(b) ..........................................................................................

(c) ..........................................................................................

(d) ..........................................................................................

(e) ..........................................................................................

(f) ..........................................................................................

(g) ..........................................................................................

(h) ..........................................................................................

(i) ..........................................................................................

(j) ..........................................................................................

(k) ..........................................................................................

(l) ..........................................................................................

## Terms relating to hatchways

*Study the diagram below showing the layout of a traditional hatchway and its equipment:*

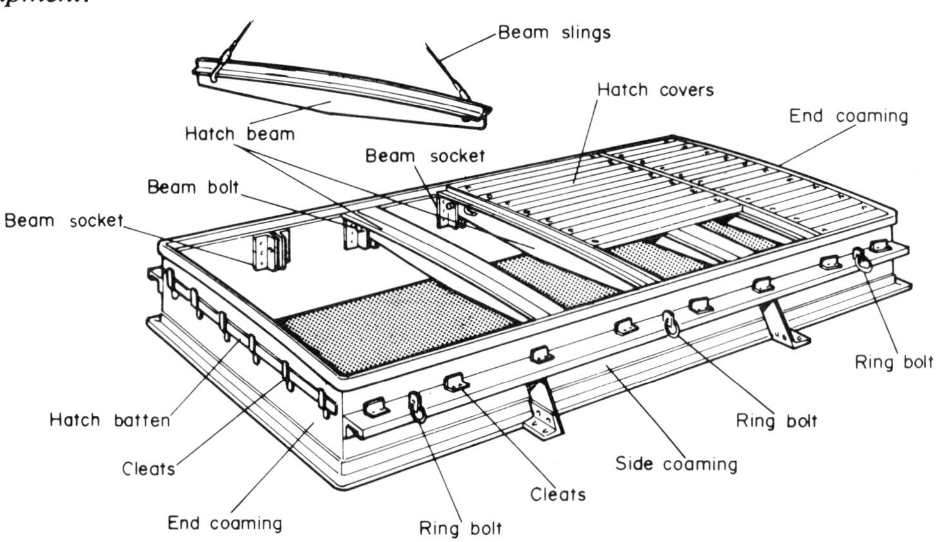

*Exercise* 2. *When the hatch is being covered, in what order would you expect the following pieces to be put in place. Use the diagram to help you.*

(a) Hatch battens (b) Hatch beams (c) Wooden wedges
(d) Hatch covers (e) Beam bolts (f) Tarpaulins

## (B) Measurement

### *(vi) Tonnage*

A ship's **underdeck, gross and net** tonnage are measurements of volume not of weight. They are measured in cubic capacity and then converted to tons using the formula 100 cubic feet = 1 ton.

**Underdeck tonnage**
This is the total volume of a ship below the tonnage deck. If the ship has double-bottom tanks, these are not included.

**Gross register tonnage**
This is the sum of the underdeck tonnage and the permanently closed-in spaces above the tonnage deck. Certain spaces are excluded. Passenger ships are usually measured in gross tons.

**Net register tonnage**
This is the cargo-carrying capacity of a ship. It can be calculated by deducting the machinery, navigating, boiler and bunker space and the crew and storage spaces from the gross tonnage. Port and canal charges are calculated on the net tonnage.

*Exercise* 1. *Label these diagrams with* ***underdeck tonnage, gross register tonnage*** *and* ***net register tonnage,*** *as appropriate:*

(a)

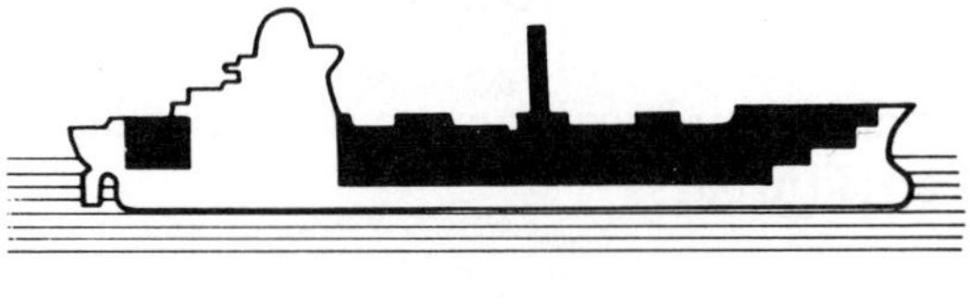

(b)

(c)

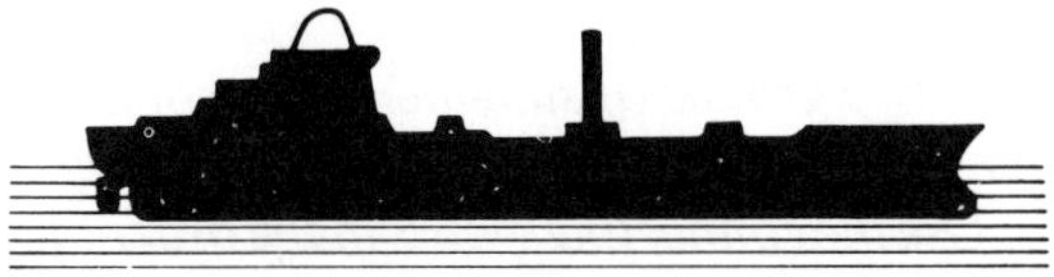

# GUIDED WRITING

## (A) Description of general cargo

*Write a description of general cargo describing how it can be classified according to its characteristics.*

*Stage* 1. *Read through the following groups of statements and join them together with the connecting words above. Each group will form a short paragraph:*

a) **which**
General cargo can be classified according to its characteristics.
General cargo is made up of an infinite variety of goods.
**however**
For example, it can be described as wet, dry, clean and dirty.
Each cargo can have more than one of these characteristics.
**because**
General cargo has different characteristics and forms of packaging.
Stowage can be a problem.

b) **which**
Wet cargoes are carried in drums and barrels.
They include such cargoes as vegetable oil and wine.
**therefore**
There is always the possibility of leakage.
They must not be stowed on top of other cargoes.

c) **but/therefore**
Dry cargoes cannot possibly leak.
They can be damaged by leakage from wet cargoes.
They must be stowed away from them.
**which**
Flour, animal feed and rice are examples of dry cargoes.
They are carried in bags.

d) **which**
Some cargoes are known as dirty cargoes.
These cargoes leave a residue behind them.

**because**
The dust from cement and fertilizers can damage other cargoes.
Holds must be cleaned out carefully after them.

*With the next two groups of statements the connecting words are given at the beginning. You must decide how many sentences to divide the statements into. Use the connecting words in the order they are given.*

e) **which/therefore**
Odorous cargoes give off fumes.
Odorous cargoes include rubber and wool.
These fumes can damage delicate cargoes.
They must be stowed separately.

f) **because/when/as**
Heavy cargoes must be given bottom stowage.
They will break more fragile cargoes.
Very heavy items are carried in ships.
They should be stowed over bulkheads.
They might damage the deck plating.

g) *With these two groups you must choose your own connecting words.*
Some cargoes are classified as dangerous cargoes.
These cargoes are combustible, toxic or liable to explode.
They are a risk to human life and the safety of the ship.
They are subject to strict laws.

h) Perishable food is usually carried in refrigerated ships.
Some general cargo vessels have refrigerated holds.
Meat and fish are often frozen solid.
Fruit and vegetables are only chilled.

*Stage 2.* *Put all the paragraphs together to form your description.*

# NOTE-TAKING PRACTICE

## Main points

As we have already seen in previous units, a lecturer will often say in his introduction (a) what his lecture is going to be about, and (b) how he is going to deal with his subject. This will help the note-taker to establish the basic outline for his notes.

## **Lecture:** Advantages and disadvantages of containerization

*Listen to the introduction to the lecture and write down a suitable title and then the main headings and sub-headings to give the basic outline for your notes.*

When a lecturer discusses a subject he will make a point and then expand on this. It is important to spot the main points and note them down.

*ıgh Part 1 of the lecture (below), which is about the advantages of container- ith reference to cost. Underline the sentences or the parts of the sentences ve five ways in which cost is reduced.*

Firstly, with reference to cost. Containerization reduces handling. This is because the goods are only packed once. They are put into the containers at the warehouse or factory and only taken out at their final destination. In addition, few people are needed to load containers into ships. All this reduces transport costs. Less handling also leads to less damage. Inside the container, which is strongly built, the goods are protected. The container also protects the goods from pilfering. That is, from being stolen. Stealing from cases which have been broken open is a serious problem in many ports around the world. A container can be securely fastened. It is hoped that less damage and less pilfering may lead to lower insurance premiums. However, this is still under negotiation with insurance companies. Another way in which containers have reduced costs is by making storage less of a problem. For example, containers waiting for shipment can be left on the quayside rather than be put in warehouses.

Having spotted the main point, it is then necessary to note it down as briefly as possible. This can be done by using abbreviations and by using the compressed form of writing practised in Unit IV.

*Try to reduce these points (the first has been done for you):*

| | | |
|---|---|---|
| (i) | containerization reduces handling | less handling |
| (ii) | less handling also leads to less damage | ........................ |
| (iii) | the container also protects the goods from pilfering | ........................ |
| (iv) | may lead to lower insurance premiums | ........................ |
| (v) | storage less of a problem | ........................ |

After making a point, a lecturer will expand it. He may add further information or examples, he may add reasons or qualifications. He will introduce these by using the connecting words and phrases studied in Units I and II (e.g. **in addition, for example, because, however**). The note-taker will introduce these by using symbols and abbreviations (e.g. **+ , &, e.g., ∵ , but**).

*Read through Part 1 again. Add more information to the main point by completing the outline below. Use the symbols to help you.*

**Title:** Adv & Disadv of Contnrztn

(A) Adv

I. With ref to cost

(i) less handling

∵ .........................

\+ .........................

(ii) less damage

∵ .........................

(iii) less pilfering (i.e. ............)

∵ .........................

(iv) lower insurance (..)

BUT.........................

(v) easier storage

e.g. .........................

*Now listen to Part 2 of the lecture, which looks at some of the advantages with reference to time. Take down notes on the outline below. Remember to listen for the main point and to note this down first. Then try to add some more information.*

*II. With ref to* .....................................

(i) .........................

∵ .........................

(ii) easier ................

∵ .........................

(iii) ........................

special ............. used

(iv) ........................

∵ ........................

*Now listen to Parts 3 and 4 of the lecture. These deal with the disadvantages of containerization. Continue taking notes for this part of the lecture.*

# GUIDED WRITING

## (B) Advantages and disadvantages of containerization

*Use the notes you have taken to write an essay on the advantages and disadvantages of containerization as a method of transporting goods. When you have finished, turn to Appendix II and compare your essay with the tapescript.*

# UNIT VII

# NAVIGATION

## READING COMPREHENSION

### (A) Navigating techniques and instruments

As we have already seen, the introduction to a reading passage will usually tell you what that passage is going to be about. You may already know from the title of the book or unit that the passage is going to be, for example, about navigation. But you will not know what aspect of navigation until you have read the introduction. This information will help your comprehension by focussing your attention on the main purpose of the passage. Your purpose for reading will be to extract that information.

*Now read just the introduction to the passage below. What aspects of navigation is the rest of the passage going to be about?*

It is the Deck Officer's job to take the ship safely from place to place as quickly and economically as possible. To do this he must have a thorough knowledge of navigation. From the moment a ship leaves a berth to her arrival at the next port, her position is constantly checked and plotted on a chart. To do this accurately, the navigator uses a number of instruments and techniques.

*Read through the rest of the passage and note down* (1) *the different techniques and* (2) *the instruments used in navigation mentioned in the passage.*

When out of sight of land, a ship's position can be found by using the techniques of celestial navigation. Celestial navigation involves taking observations of the sun, moon and stars with a sextant. This is an instrument which measures the angle between the celestial body and the horizon. The exact time that the sight is made must also be recorded. This is done on the ship's chronometer, which is a very accurate clock. With this information and the tables given in a book called a nautical almanac, the navigator is able to calculate the ship's position. The position is marked on the chart by a dot with a circle round it. A time is also given. By joining up the dots with a pencil line, the ship's track can be seen.

A gyro compass (*Unit VII*)

A magnetic compass (*Unit VII*)

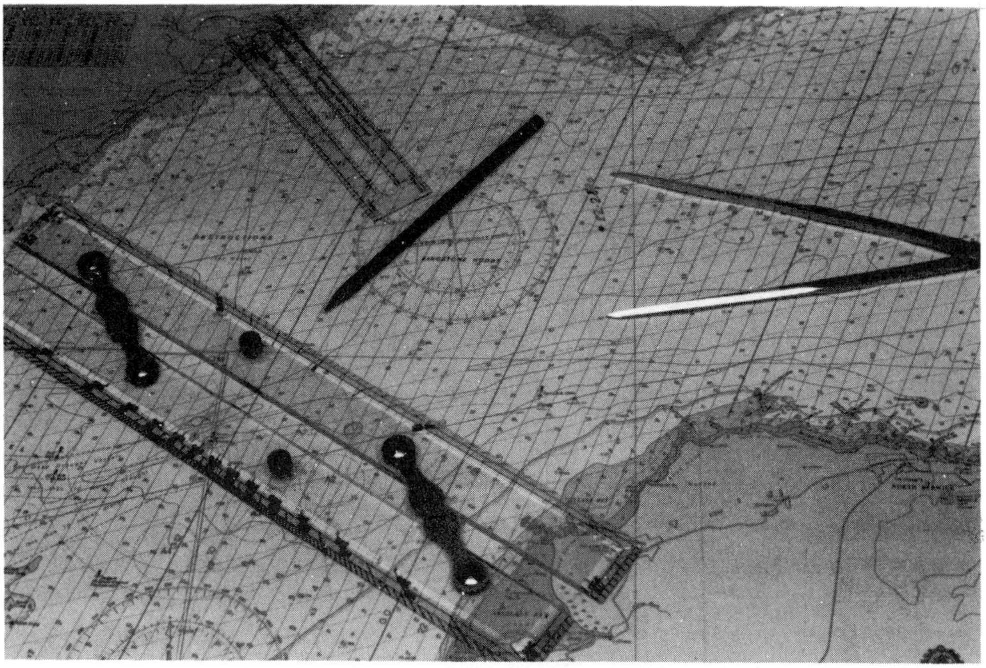

Navigating instruments (*Unit VII*)

A sextant (*Unit VII*)

While in sight of land, the navigator uses the techniques of coastal navigation to find his position. Navigation in coastal waters is known as pilotage. At regular intervals the officer on watch takes observations of conspicuous landmarks, such as a lighthouse or a church spire, using the ship's compass and a sextant. By looking at the chart to find out the height of the object and by making adjustments for tide and the height of the bridge from sea level, he can calculate the distance of the ship from the object. The ship's position will now lie somewhere on the circumference of a circle with the object at its centre and the distance off as the radius. To find the exact position of the ship a compass bearing can be used. A position line showing the bearing of the ship from the object is drawn on the chart. The position of the ship is where the line cuts the circle. This is known as making a fix. There are other ways of making a fix. For example, two, or better still, three simultaneous compass bearings can be taken of different objects. The objects should not be too close together. Theoretically the lines from these objects should intersect at the same point. In practice they usually form a triangle, which is called a cocked hat. If this is small, the ship's position is assumed to be at the point of the triangle nearest any danger. If the triangle is large, the bearings should be taken again.

A fairly accurate estimation of a ship's position can be calculated by a technique known as dead reckoning (DR). To do this the navigator needs to know the ship's course, speed and distance run. A DR position is made more difficult to calculate by the effects of wind, tide and current.

*Draw two sketches to illustrate the two ways of making a fix described in the paragraph on pilotage.*

# GRAMMAR

## (A) Prepositions of place (ii)

*Study how the following prepositions are used to describe destination:*

1. **To/from** – used with reference to a point

| | |
|---|---|
| **to → ●** | **● → from** |
| They took the ferry to Calais. | The QE II sailed from Southampton. |

2. **Onto/off** – used with reference to a line or surface

| | |
|---|---|
| **onto ↘___** | **___↗ off** |
| The case fell onto the deck. | The launch took off the pilot. |

3. **Into/out of** – used with reference to an area or volume

**into** 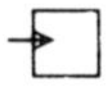

The cargo was lowered into the hold.

**out of** 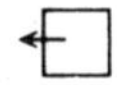

The cargo was lifted out of the hold.

*Now study how these are used to describe direction:*

4. **Towards/away from** – express motion with reference to a directional path

**towards** ----→ ●

The ship was driven towards the rocks.

● ----→ **away from**

The wood drifted away from the shore.

5. **Up/down** – express motion with reference to a vertical axis.

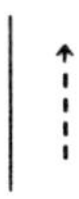 **up**

The sailor climbed up the mast.

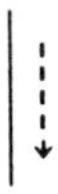 **down**

The flag was hauled down.

6. **Across/along** – express motion with reference to a horizontal axis or plane

**across** 

The dinghy tacked across the bay.

**along** 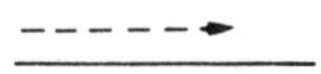

They sailed along the coast.

*Now study how these are used to describe passage:*

7. **over**

The people walked over the bridge.

**under**

The barge sailed under the bridge.

8. **behind**

The plane passed behind the lighthouse.

**in front of**

The gulls flew in front of the lighthouse.

9. **(a)round**

The diver swam round the shoal.

**through**

The diver swam through the shoal.

**10. by, past**

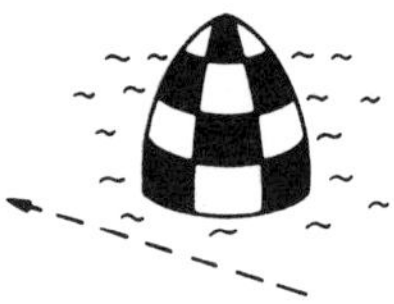

The sailors rowed past/by the buoy.

*Exercise* 1. *Read this description of a voyage made by the MV* Diomede. *Then write similar descriptions of the other two voyages outlined below. Use the appropriate prepositions and the definite article (the) where necessary:*

On 6 April, 1979 the general cargo vessel, the MV *Diomede,* left the port of New Orleans in the USA for Hawaii in the Hawaiian Islands. She steamed across the Gulf of Mexico, through the Yucatan Channel and along the north coast of Jamaica, calling at Kingston. Then she sailed through the Panama Canal and across the Pacific Ocean to Hawaii.

10/9/80 – bulk oil carrier – SS *Enterprise* – Nagasaki – Japan – Kharg Island – Gulf – east coast of China – Taiwan Straits – South China Sea – Singapore – Straits of Malacca – Indian Ocean – Kharg Island.

16/6/80 – passenger liner – MV *Orient* – Southampton – UK – Naples – Italy – English Channel – Bay of Biscay – west coast of Portugal – Lisbon – Straits of Gibraltar – Mediterranean Sea – Naples.

*Exercise* 2. *Write a similar description of any voyage made by your ship.*

## (B) Relative clauses (ii)

1. Reduced relative clauses

This is possible:
When the *defining* relative clause contains a verb in the passive
e.g. A ship which was built in that yard sank yesterday.
⇒ A ship built in that yard sank yesterday.

When the *defining* relative clause contains a verb in a continuous tense
e.g. A ship which is carrying explosives flies a red flag.
⇒ A ship carrying explosives flies a red flag.

When the *defining* relative clause begins: **which has/have**
e.g. Vessels which have 'tween decks are suitable for general cargo.
⇒ Vessels having 'tween decks are suitable for general cargo.
**or** Vessels with 'tween decks are suitable for general cargo.

When a *defining* relative clause contains a verb in the simple present.
e.g. Drums which contain chemicals are sometimes stowed on deck.
⇒ Drums containing chemicals are sometimes stowed on deck.

*Exercise* 1. *Reduce the following relative clauses if they can be reduced:*

(a) Ships which use the Panama Canal pay canal dues.
(b) Rope which is made of nylon does not rot.
(c) An officer who has a Master's Certificate can hope to become the master of his own ship one day.
(d) Some boats are built of fibre glass, which is a type of plastic.
(e) A sailor who is working aloft should wear a safety harness.
(f) Steel which contains more than 0.7 per cent of carbon has a high tensile strength.
(g) Cargoes which have a low stowage factor are usually put at the bottom of the hold.
(h) A material which can be shaped by hammering or rolling is said to be malleable.
(i) Ships which are classed by Lloyds have to undergo strict tests.
(j) The sailors who were sitting at the back of the coach were hurt in the accident.

2. Relative clauses with prepositions

*Study how these two sentences are joined together:*

(a) Containers are large cases of standard dimensions.
(b) Cargo is packed in these.
(a) + (b) Containers are large cases of standard dimensions in which cargo is packed.
(*Note:* In spoken English the preposition is usually put at the end.)

*Exercise* 2. *Join the following pairs of sentences together in the same way by making the second sentence a relative clause:*

(a) Copper and tin are non-ferrous metals.
Bronze is made from copper and tin.
(b) Pallets are pieces of cargo-handling equipment.
Crates and cases are stacked on pallets.
(c) The windlass is found on the forecastle.
The anchors are raised and lowered by the windlass.

(d) Ventilators are positioned over the cargo holds.
Air flows through these.

(e) Hatchways are covered by hatch covers.
Tarpaulins are then spread over the hatch covers.

(f) A sextant is a navigating instrument.
With a sextant a Deck Officer takes observations of the sun.

(g) Speed trials take place when the ship is out at sea.
During speed trials the ship's maximum speed is measured.

(h) The National Union of Seamen is an organization which looks after the interests of seamen.
All Merchant Navy ratings must belong to the Union.

(i) The flight deck is an open deck on an aircraft carrier.
From this deck planes and helicopters take off.

# READING COMPREHENSION

## (B) Electronic aids to navigation

*The following passage is a continuation of the reading passage in Reading Comprehension (A). Read through the introduction to find out what the rest of the passage is going to be about.*

*Now read through the rest and make a list of the electronic devices used for navigation.*

In addition to the traditional methods of navigation a number of sophisticated electronic devices have been developed to help the navigator. These include the echo sounder, radar and various types of radio navigation equipment.

The echo sounder sends a radio signal from the bottom of the ship to the sea-bed, from which it is reflected. The time taken to receive the reflected signal is a measure of the depth of water under the ship. The received pulse is displayed on a chart by a pen recorder so that the navigator can see the outline of the bottom over which the vessel is passing. A similar device is the sonar system, which uses high frequency sound signals. In sonar the sound signal can be sent ahead or sideways. The time for the echo to be sent back from an object, such as an underwater rock, is a measure of the object's distance from the ship. The sonar system can also be used to measure the speed of the ship over the seabed.

It is now common for ships, including trawlers, tugs and ferries, to be fitted with radar to help them navigate near land. High-frequency radio beams are sent out and received by a rotating radar scanner on the ship. These signals are then converted into images on a screen. The ship is shown in the centre of the screen and nearby ships are shown in relation to it. On the screen moving objects leave a short trail to indicate their direction and speed. But radar is only an aid to navigation. It is the correct interpretation of the picture on the screen which ensures the safe passage of the ship.

Nearly all seagoing vessels are fitted with radio for transmitting and receiving signals. One of the most common methods of checking a vessel's position is by using the radio direction finder. RDF is a simple system in which the operator finds the direction of two homing beacons. However, the system is often not very accurate.

Equipment such as the Decca Navigator, the Omega Navigator and Loran, use synchronized signals received from specific shore stations to establish position. This may be done either by comparing the time of arrival of a certain signal from three or more stations, or by comparing the difference in phase of radio waves. They all work on the same principle. Signals received from a pair of stations place the ship on a curved line of position. Anywhere along this line the received signals from the two stations would seem the same. Signals received from another pair of stations define a second line of position. The ship's position is where the two lines cross. The position is displayed on a chart and on dials. The accuracy of this system is usually within half a mile.

Since the late 1960s it has been possible to navigate by satellite. Signals from satellites are received by a dish-shaped aerial on the ship. As the satellite approaches or goes away from the ship, the character of the signal received changes. This change depends on where the ship is in relation to the orbiting satellite. Because the position of the satellite is known with very high accuracy the ship's position can also be fixed. This system gives an accuracy of better than one tenth of a mile in all weathers, all over the world. The introduction of satellites has led to fully automatic navigation in the open sea for modern ships.

The ultimate in navigation may eventually be in a dual system combining satellites with inertial navigation. In inertial navigation, sensors measure the ship's speed and direction, and compute the changing position from these readings. This system is entirely internal and independent of any external references such as landmarks, the sun, stars, radio or radar transmission. The equipment only needs to know the exact location of the starting point. Using this system, which is extremely accurate, a ship can be set on course and kept on it despite current, winds and alterations in engine power. Inertial navigation is used in warships.

# APPLIED TERMINOLOGY

## (A) Terms relating to compass cards

*Study the compass card below. It is divided up clockwise into 360 degrees. It also shows eight of the points of the compass (N, NE, E, etc.).*

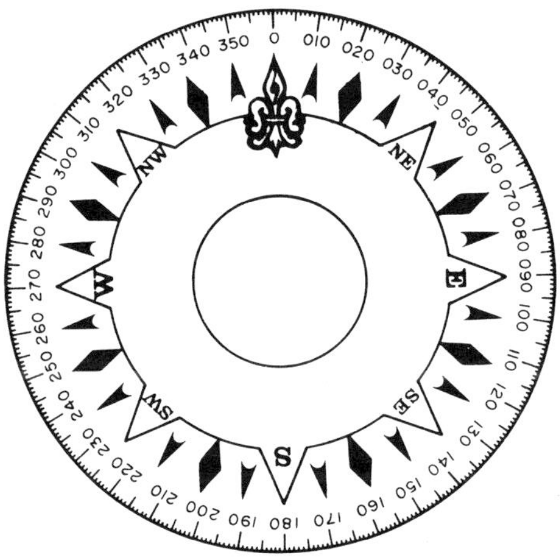


*Exercise* 1. *Write out in full the eight points of the compass shown above (the first two have been done for you):*

(a) N = North
(b) NE = North-East
(c) E = ..............
(d) SE = ..............
(e) S = ..............
(f) SW = ..............
(g) W = ..............
(h) NW = ..............

*Exercise* 2. *What would the following points be in degrees?*

(a) S
(b) NE
(c) NW
(d) E
(e) SW
(f) N
(g) W
(h) SE

# Terms describing position in relation to a ship

*Study the diagram below*

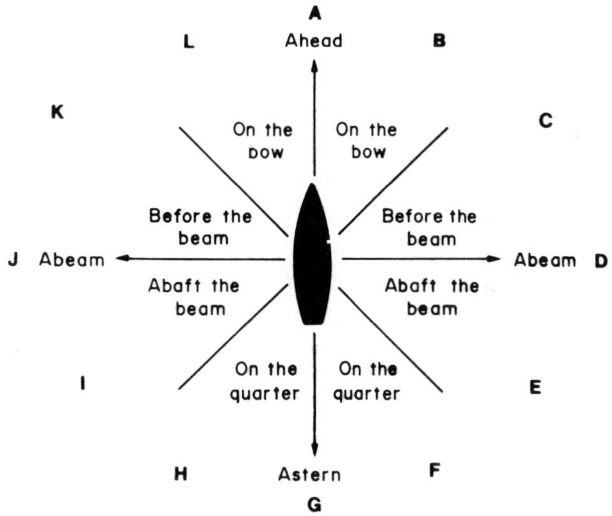


In order to describe the position of ships in relation to the tanker, they can be used as follows:

1. Ship A is (dead) ahead./Ship A is ahead of the tanker.
2. Ship B is on the starboard bow.
3. Ship C is before the starboard beam.
4. Ship D is abeam./Ship D is on the starboard beam.

*Exercise* 3. *Now continue describing the position of the other ships:*

(a) Ship E is ..................................................................

(b) Ship F is ..................................................................

(c) Ship G is ..................................................................

(d) Ship H is ..................................................................

(e) Ship I is ..................................................................

(f) Ship J is ..................................................................

(g) Ship K is ..................................................................

(h) Ship L is ..................................................................
(*Note:* 'dead' in example 1 means directly ahead.)

# (B) Measurement

## (vii) Distance, speed

1. Nautical miles

Distance at sea is measured in nautical miles. 1 nautical mile is the length of an arc of a great circle which subtends an angle of 1 minute at the centre of the Earth. 1 nautical mile = 1.151 statute miles = 1.853 kilometres.)

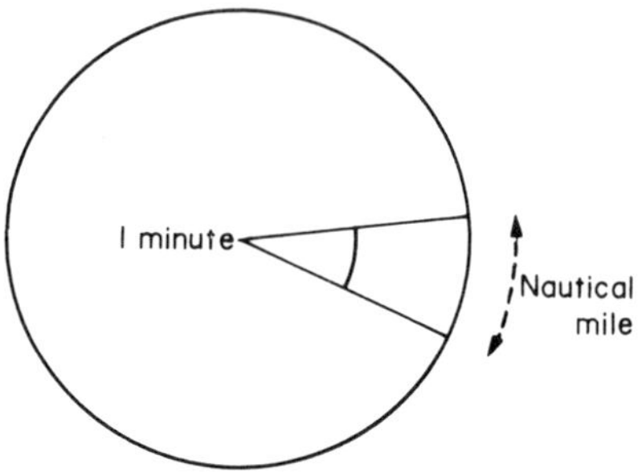


(Not drawn to scale)

*Exercise* 1. *Convert the following angular distances into nautical miles:*

| | | |
|---|---|---|
| (a) 15° 36′ | (b) 42° 17′ | (c) 35° 35′ |
| (d) 21° 59′ | (e) 32° 30′ | (f) 45° 19′ |

2. Knots

At sea, speed is measured in knots. (1 knot = 1 nautical mile per hour.)

*Exercise* 2. *Complete this chart, using the formula:* $speed = \frac{distance}{time}$

| MV *Panama* | Distance | Time | Speed |
|---|---|---|---|
| Monday | 180 N Miles | 12 hours | ........................ |
| Tuesday | ........................ | 24 hours | 14 knots |
| Wednesday | 288 N Miles | 24 hours | ........................ |
| Thursday | ........................ | 8 hours | 6.5 knots |
| Friday | ........................ | 10 hours | 11 knots |
| Saturday | 312 N Miles | 24 hours | ........................ |
| Sunday | 152 N Miles | 16 hours | ........................ |

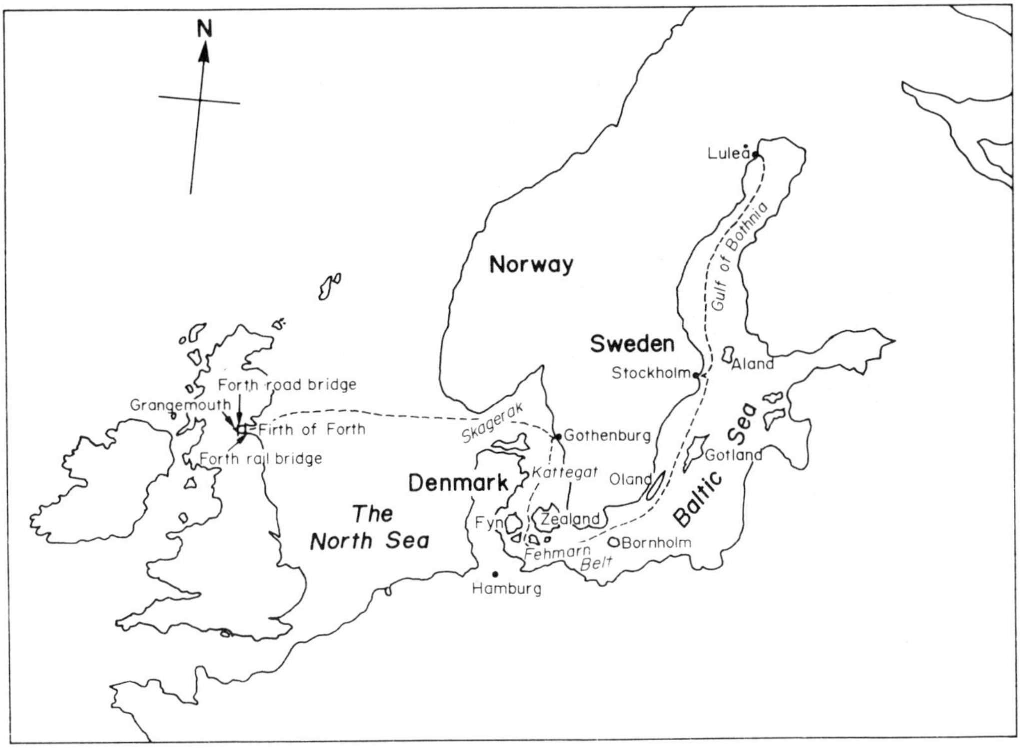


# GUIDED WRITING

## (A) Description of a voyage

*Write a description of a voyage made by the MV* Voyager *from the port of Grangemouth in Scotland to the ports of Gothenburg, Stockholm and Luleå in Sweden.*

*Stage* 1. *Study the route shown on the map opposite.*

*Stage* 2. *Read through the timetable below and the information in the statements:*

| | |
|---|---|
| 10 May sailed from Grangemouth | 15 May arrived at Stockholm |
| 11 May arrived at Gothenburg | 16 May sailed from Stockholm |
| 12 May sailed from Gothenburg | 19 May arrived at Luleå |

(a) The MV *Voyager* was an oil tanker.
(b) She was carrying a cargo of petroleum products.
(c) Grangemouth is a small port on the River Forth.
(d) Near the port there are important oil refineries.
(e) Her passage across the North Sea was good.
(f) The weather was clear and the wind was light.
(g) Gothenburg is a port on the west coast of Sweden.
(h) At Gothenburg she discharged gas oil.
(i) At Gothenburg she loaded petrol.
(j) Stockholm is the capital of Sweden.
(k) She unloaded more of her cargo at Stockholm.
(l) Luleå is only open during the summer months.
(m) From Luleå iron-ore is exported.
(n) She unloaded the rest of her cargo at Luleå.
(o) She set sail for Grangemouth.
(p) On her way through the Kattegat she was ordered to proceed to Hamburg.
(q) Hamburg is a port in Germany.

*Stage* 3. *Now write your description using all the information available. When appropriate use relative clauses and the connecting words that you have learned so far.*

# NOTE-TAKING PRACTICE

## Specialized abbreviations and symbols

## **Lecture:** Shipping forecasts

*Shipping forecasts are broadcast by radio stations throughout the world at set times during the day. A shipping forecast will give you details of the weather conditions in your areas and contains a lot of useful information which is difficult to note down without a system of symbols and abbreviations and a carefully organized outline to guide you. For example listen to the beginning of the forecast for the sea area around the British Isles, as might be broadcast by the BBC, and try to note down the information before it fades out. Now study the map and the completed table on page 113 while listening to the complete forecast. The following notes explain how the information is noted in the table:*

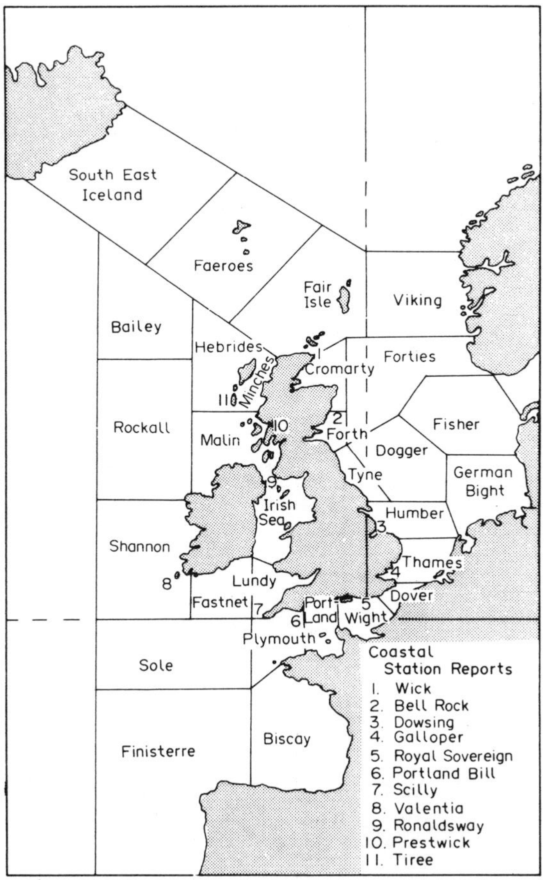


1. Write down the date and time of the broadcast at the top, i.e. 24 May 0640.
2. Put a tick (√) in the column marked 'gales' against the places where there are gales, i.e. Dogger, Fisher, German Bight.
3. Write down the time of the general synopsis, i.e. 0000.
4. Write down the information relating to the pressure systems. Do not write in the words 'millibars'. Use abbreviations for North, South etc., i.e. N, S. The following

Weather Forecast Date: 24th May Time: 0640 hrs.

General Synopsis at 0000 G.M.T.

| System | Present Position | Movement | Forecast Position | At |
|---|---|---|---|---|
| L 988 | 60N 10E | NE | | |
| | | | | |
| L 1004 | 56N 19W | E | S SCOT | 2400 |
| | | | | |

Forecast for Sea Areas

| Area | Gales | Wind | | Weather | Visibility |
|---|---|---|---|---|---|
| | | Now | Later | | |
| Viking | | NW6 | NW5/6 | pr | m/g |
| Forties | ✓ | NW6/8 | NW3/5 | pr | m/g |
| Cromarty | | NW3/5 | SE 6 | r | g/m |
| Forth | | | | | |
| Tyne | | | | | |
| Dogger | ✓ | | | | |
| Fisher | ✓ | NW6/8 | NW4/5 | pr | g |
| German Bight | ✓ | | | | |
| Humber | | W3/5 | | | |
| Thames | | | SW5/7 | r | g |
| Smith's Knoll | | | | | |
| Dover | | W 6 | | | |
| Wight | | | | | |
| Portland | | W4/6 | | r/d/m | p |
| Plymonth | | | | | |
| Biscay | | W 3 | | c | g |
| Finisterre | | NE3/5 | | c | g |
| Sole | | | | | |
| Lundy | | | | | |
| Fastnet | | SW 4/6 | | r/d/m | g/p |
| Irish Sea | | | | | |
| Shannon | | | | | |
| Rockall | | SW4/6(S) SE3/5(N) | | | |
| Malin | | Var 3 | S/7 at centre of L | | |
| Hebrides | | Var | E 3/6 | c | g |
| Minches | | | | | |
| Fair Isle | | N6 | E6 | c | g |
| Bailey | | SE5 | | c | g |
| Faeroes | | N3/6 | E3/5 | c | g |
| S.E. Iceland | | Var3 | SE 4/6 | c | g |

Reports from Coastal Stations at 0400 hrs. G.M.T.

| Station | Wind | | Weather | Visibility | Barometer | Tendency |
|---|---|---|---|---|---|---|
| | Direction | Force | | | | |
| Wick | W | 2 | | 16 | 1016 | / |
| Bell Rock | W | 5 | c | 22 | 1016 | — |
| Dowsing | WNW | 4 | c | 5 | 1019 | |
| Galloper | W | 4 | | 5 | 1021 | — |
| Royal Sovereign | W | 4 | | 16 | 1022 | — |
| Portland Bill | W×S | 3 | | 22 | 1022 | \ |
| Scilly | SW | 4 | r | 16 | 1023 | \ |
| Valentia | SW | 4 | r | 6 | 1017 | \ |
| Ronaldsway | Calm | | r | 9 | 1017 | \ |
| Prestwick | SE | 2 | r | 13 | 1017 | \ |
| Tiree | ESE | 4 | | 27 | 1015 | \ |

abbreviations can be put in front of the figure for millibars:

L = low pressure, depression
H = high pressure, anticyclone
T = trough of low pressure
R = ridge of high pressure
W = warm front
C = cold front
O = occluded front
S = secondary front

5. Put a bracket round the sea areas given together, e.g. Dogger, Fisher, German Bight.

6. Fill in the column for wind. A stroke (/) indicates an increase or decrease in the wind's force. For example:
   Wind north-westerly force 6 to 4 = NW 6/4.

7. Fill in the column marked weather using these abbreviations:

| **Weather** | **Letter** | **Weather** | **Letter** |
|---|---|---|---|
| blue sky | b | overcast sky | o |
| partly cloudy sky | bc | squally weather | q |
| cloudy | c | rain | r |
| drizzle | d | sleet | rs |
| fog | f | snow | s |
| gale | g | thunder | t |
| hail | h | thunderstorm | tlr |
| lightning | l | dust haze | z |
| mist | m | | |

(*Note:* (a) 'p' denotes 'showers of' e.g. pr = showers of rain
(b) capital letters denote intensity e.g. R = heavy rain
(c) letters repeated denote continuity e.g. RR = continuous heavy rain
(d) 'i' in front denotes intermittent e.g. ir = intermittent rain
(e) a stroke (/) divides present weather from future weather
e.g. r/d/m = rain followed by drizzle then mist.)

8. Fill in the column marked 'visibility' using the following abbreviations:
   f = fog  m = moderate  vg = very good
   p = poor  g = good

9. Fill in the time of the reports from coastal stations.

10. Fill in the column using abbreviations and figures as appropriate.
    dir'n = direction  Vis = visibility  Baro = barometer reading

11. Fill in the column marked 'tendency' as follows:
    / for a rising pressure, — for a steady pressure and \ for a falling pressure.
    The angle of the sloping line equals the rate of rise or fall in pressure.

*When you are thoroughly familiar with the symbols and abbreviations, listen to the same forecast again and try and complete the table yourself, without looking at the completed version. Check your notes against the completed version afterwards.*

Weather Forecast Date: Time:

General Synopsis at G.M.T.

| System | Present Position | Movement | Forecast Position | At |
|---|---|---|---|---|
| | | | | |
| | | | | |
| | | | | |
| | | | | |

Forecast for Sea Areas

| Area | Gales | Wind | | Weather | Visibility |
|---|---|---|---|---|---|
| | | Now | Later | | |
| Viking | | | | | |
| Forties | | | | | |
| Cromarty | | | | | |
| Forth | | | | | |
| Tyne | | | | | |
| Dogger | | | | | |
| Fisher | | | | | |
| German Bight | | | | | |
| Humber | | | | | |
| Thames | | | | | |
| Smith's Knoll | | | | | |
| Dover | | | | | |
| Wight | | | | | |
| Portland | | | | | |
| Plymonth | | | | | |
| Biscay | | | | | |
| Finisterre | | | | | |
| Sole | | | | | |
| Lundy | | | | | |
| Fastnet | | | | | |
| Irish Sea | | | | | |
| Shannon | | | | | |
| Rockall | | | | | |
| Malin | | | | | |
| Hebrides | | | | | |
| Minches | | | | | |
| Fair Isle | | | | | |
| Bailey | | | | | |
| Faeroes | | | | | |
| S.E. Iceland | | | | | |

Reports from Coastal Stations at G.M.T.

| Station | Wind | | Weather | Visibility | Barometer | Tendency |
|---|---|---|---|---|---|---|
| | Direction | Force | | | | |
| Wick | | | | | | |
| Bell Rock | | | | | | |
| Dowsing | | | | | | |
| Galloper | | | | | | |
| Royal Sovereign | | | | | | |
| Portland Bill | | | | | | |
| Scilly | | | | | | |
| Valentia | | | | | | |
| Ronaldsway | | | | | | |
| Prestwick | | | | | | |
| Tiree | | | | | | |

# GUIDED WRITING

## (B) Description of weather conditions

*Write a description of the weather conditions in your area today.*

*Give a time and date and the name of the area in which the conditions occur. Give a full description of wind direction and speed, temperatures, clouds, visibility, air pressure and general weather conditions.*

Information may be got from the daily newspapers, the weather forecast on the radio and television, nearby meteorological instruments, by telephoning your local weather centre and by your own observations.

# UNIT VIII

# MAIN ENGINES

## READING COMPREHENSION

### (A) Different types of marine engine

*Read through the following passage on the different types of marine engine. Use the techniques that you have practised so far to help you.*

There are four main types of marine engine: the diesel engine, the steam turbine, the gas turbine and the marine nuclear plant. Each type of engine has its own particular application.

The diesel engine is a form of internal combustion engine similar to that used in a bus. Its power is expressed as brake horsepower (bhp). This is the power put out by the engine. Effective horsepower is the power developed by the piston in the cylinder, but some of this is lost by friction within the engine. The power output of a modern marine diesel engine is about 40,000 brake horsepower. This is now expressed in kilowatts. By comparison the engine of a small family car has an output of about 80 bhp. Large diesel engines, which have cylinders nearly 3 ft in diameter, turn at the relatively slow speed of about 108 rpm. These are known as slow-speed diesel engines. They can be connected directly to the propeller without gearing. Although higher power could be produced by higher revolutions, this would reduce the efficiency of the propeller, because a propeller is more efficient the larger it is and the slower it turns. These large slow running engines are used in the larger merchant ships, particularly in tankers and bulk carriers. The main reason is their low fuel consumption. More and more of the larger merchant vessels are being powered by medium-speed diesel engines. These operate between 150 and 450 rpm, therefore they are connected to the propeller by gearing. This type of engine was once restricted to smaller cargo ships, but now they are used in fast cargo liners as well as in tankers and bulk carriers. They are cheaper than slow-speed diesel engines, and their smaller size and weight can result in a smaller, cheaper ship.

In steam turbines high pressure steam is directed into a series of blades or vanes attached to a shaft, causing it to rotate. This rotary motion is transferred to the propeller shaft by gears. Steam is produced by boiling water in a boiler, which is fired by oil. Recent developments in steam turbines which have reduced fuel consumption and raised power output have made them more attractive as an alternative to diesel power in ships. They are 50 per cent lighter and on very large tankers some of the steam can be used to drive the large cargo oil pumps. Turbines are often used in container ships, which travel at high speeds.

Gas turbines differ from steam turbines in that gas rather than steam is used to turn a shaft. These have also become more suitable for use in ships. Many naval vessels are powered by gas turbines and several container ships are fitted with them. A gas turbine engine is very light and easily removed for maintenance. It is also suitable for complete automation.

Nuclear power in ships has mainly been confined to naval vessels, particularly submarines. But this form of power will be used more in merchant ships as oil fuels become more expensive. A nuclear-powered ship differs from a conventional turbine ship in that it uses the energy released by the decay of radioactive fuel to generate steam. The steam is used to turn a shaft via a turbine in the conventional way.

*Complete this diagram to form a summary of the basic types of marine engine.*

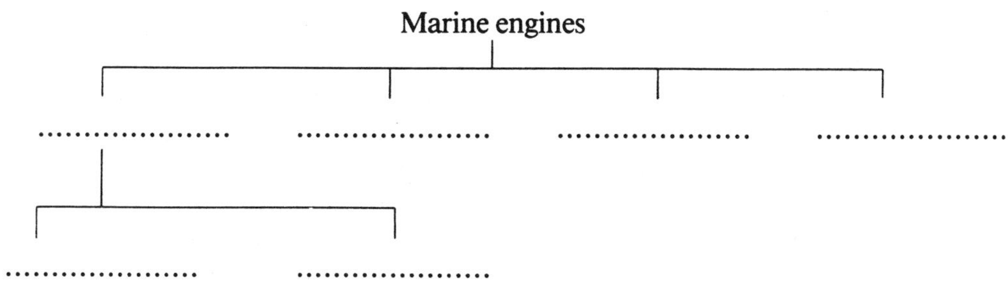


*Now fill in this box to summarize the application of these engines.*

| Type of engine | Application |
|---|---|
| slow-speed diesel | |
| medium-speed diesel | |
| steam turbine | |
| gas turbine | |
| marine nuclear plant | |

# GRAMMAR

## (A) Causal verbs

There are several verbs which can be used to express cause.

1. **To have** and **to get**

   *Study these examples:*

   He had the engines overhauled.
   He got the engines overhauled.

   In these sentences the subject 'he' caused the action, but did not perform it himself. In the following sentences the person performing the actions is introduced. Note how the infinitive is used.

   He had the Engine Department overhaul the engines.
   He had the engines overhauled by the Engine Department.
   He got the Engine Department to overhaul the engines.
   He got the engines overhauled by the Engine Department.

*Exercise* 1. *Write out these notes using the verbs in brackets and the appropriate pattern above.*

e.g. (*had, to examine*) The Master – the equipment – the electrician
The Master had the equipment examined by the electrician.

(a) (*got, to plot*) The Second Mate – the cadet – the course
(b) (*had, to check*) The First Mate – the cargo
(c) (*had, to lower*) The Bosun – the boats – the apprentices
(d) (*got, to clean*) The Carpenter – the bilges – the ratings
(e) (*got, to adjust*) The Master – the compasses
(f) (*had, to supervise*) The Chief Engineer – the Second Engineer – the work
(g) (*got, to lubricate*) The Second Engineer – the greasers – the machinery
(h) (*had, to sweep*) The Chief Steward – the saloon – the galley boy

2. **-en** Verbs

   Verbs ending or beginning with **-en** mean to make something happen. The -en is usually attached to the adjective, but sometimes the noun is used.

   e.g. To make tight = to tighten
   To make longer = to lengthen

*Exercise* 2. *Rewrite these sentences using an -en verb:*

(a) The carpenter made the plank shorter.
(b) The gap between the two vessels became wider.

(c) You need to make your pencil sharp before doing chartwork.
(d) Constant pressure makes a spring weak.
(e) Speed was reduced to make the load lighter on the engine.
(f) The ropes had to be made longer because they would not reach.
(g) He had great difficulty in making the nut loose.
(h) Adding carbon to steel makes it harder.
(i) Reducing the carbon content makes it softer.
(j) The hole had to be made larger before the bolt would fit.
(k) Friction is increased by making a surface rougher.

3. Causal verbs
   (a) General causal verbs: **to make, to cause**
   (how they are used is shown in Part (B) following)
   (b) Specific causal verbs: **to increase, to raise, to reduce, to lower,** etc.
   (see Part (B) following)
   (c) Verbs enabling something to happen: **to permit, to allow, to let**
   e.g. The piston went down, permitting/allowing air *to enter* the cylinder.
   The piston went down, letting air *enter* the cylinder.

## (B) Cause and effect

Cause and effect relationships can be linked in a number of ways.

1. When both are nouns they are linked as follows:

| **cause** | ⟶ | **effect** |
|---|---|---|
| sea water | causes | corrosion |

Other links are: results in, leads to, gives rise to, produces, is the cause of, is one of the causes of.

The effect may be put first:

| **effect** | ⟵ | **cause** |
|---|---|---|
| Corrosion | is caused by | sea water |

Other links are: results from, is due to, arises from, is the result of
(*Note:* If this is not the only cause, the linking phrase must be modified by a modal, e.g. **may/can** or a frequency adverb, e.g. **often/sometimes.**)

*Exercise* 1. *These cause and effect pairs have been mixed up. Find the corresponding parts and join them together using any of the linking phrases above as appropriate.*

| | |
|---|---|
| (a) poor lubrication | a reduction in power |
| (b) instability | fuel economy |
| (c) overloading | accidents |

| | |
|---|---|
| (d) efficiency | breakdowns |
| (e) heat loss | good management |
| (f) carelessness | structural strain |
| (g) careful navigation | bad loading |
| (h) lack of maintenance | friction |

2. When both are clauses they can be linked as follows:
**cause:** The engines reduced speed.
**effect:** The ship slowed down.

(a) by using the general causal verbs **to cause, to make**
e.g. The engines reduced speed, causing the ship *to slow* down.
The engines reduced speed, making the ship *slow* down.

(b) by using **-ing** clauses
e.g. The engines reduced speed, slowing the ship down.

(c) by using connecting words, such as **consequently**
e.g. The engines reduced speed, consequently the ship slowed down.

(d) by using specific causal verbs e.g. **to raise** (to cause to rise)
**cause:** The piston rises.
**effect:** The pressure in the cylinder rises.
e.g. The piston rises, raising the pressure in the cylinder.

*Exercise* 2. *Study this diagram of an engine pressure indicator and complete the sentences below using any of the cause and effect links studied so far.*

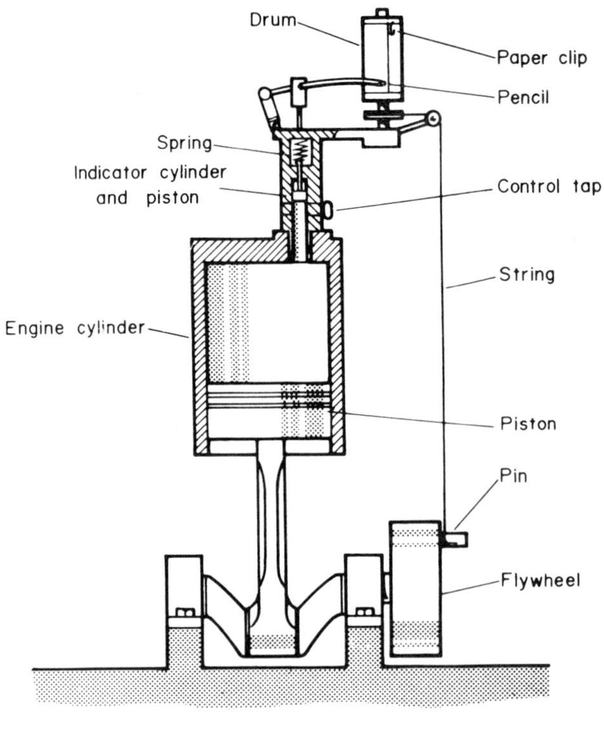

With the tap closed:

(a) The piston is driven downwards, ............... the flywheel to rotate.

(b) The flywheel rotates, ......... the pin travel downwards.

(c) The pin travels downwards, ......... the string.

(d) The string is attached to the drum in such a way that it ......... the drum to rotate.

(e) When the piston travels upwards, the tension on the string is .........,

......... the drum to return to its original position by means of a spring.

(f) Therefore each time the flywheel rotates, it ......... the drum to rotate backwards and forwards.

(g) The point of a pencil is held against the paper on the drum, ............ the pencil marks the length of the stroke on the paper each time the drum rotates.

With the tap open:

(h) The upward movement of the piston ............ compression of the gas in the cylinder.

(i) The pressure of the gas ............ the indicator piston.

(j) The indicator piston rises, ............ the spring.

(k) The piston is attached to the arm holding the pencil, ............ the pencil moves upwards ............ the rise on the paper.

# READING COMPREHENSION

## (B) Two-stroke and four-stroke cycles

Generally technical books try to give a lot of information in a short space. To do this they use a large number of diagrams. Also they often use a condensed form of writing similar to notes. For complete comprehension the reader needs to keep one eye on the diagram and the other eye on the text.

*Read through the description below of the operating cycle of a two-stroke diesel engine, using the diagram to help you. Keep reading it through until you understand it.*

A two-stroke cycle takes place in two consecutive strokes of the engine piston, or one revolution of the engine. The two strokes of the cycle are: the *compression stroke* and the *power* or *expansion stroke.* Each operation in the cycle can be shown on a time diagram. The diagram below shows a typical two-stroke cycle with the operations numbered.

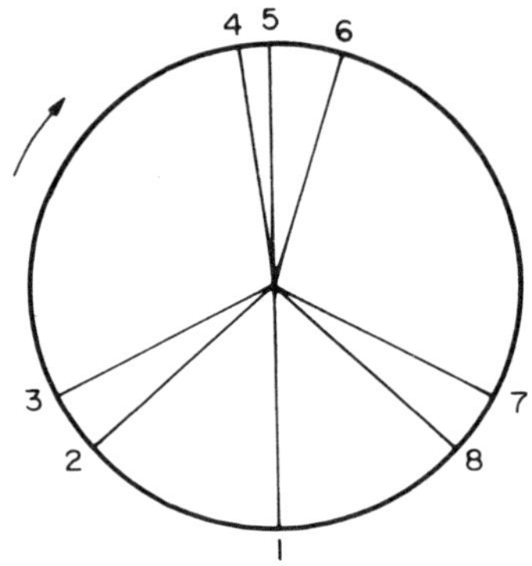


1–2 Completion of scavenge. Air enters the cylinder, expelling exhaust gas and recharging it for combustion. Scavenge and exhaust ports are open.

2–3 Post-scavenge period. Scavenge ports have closed and some air in the cylinder may leak to exhaust.

3–4 Compression. Exhaust port has now closed and the air trapped in the cylinder is compressed by the upstroke of the piston to raise its temperature sufficiently to ignite the fuel.

4–5–6 Fuel injection takes place and combustion occurs causing a rapid rise in pressure.

6–7 Expansion. Combustion completed, the hot gases expand forcing the piston downwards.

7–8 Exhaust blow down. Exhaust port has opened, allowing gas to pass to exhaust manifold and pressure drops rapidly in cylinder.

8–1 Scavenge. Scavenge ports have opened and air enters to expel remaining exhaust gas.

*Now label the diagram with* **completion of scavenge, post-scavenge, compression, fuel injection, expansion, exhaust, scavenge,** *in the appropriate places.*

The next diagram and notes illustrate a four-stroke cycle. This takes place in four strokes of the engine piston, or two revolutions of the engine. These strokes are: the *compression stroke,* the *power* or *expansion stroke,* the *exhaust stroke* and the *aspiring* or *induction stroke. Now read on.*

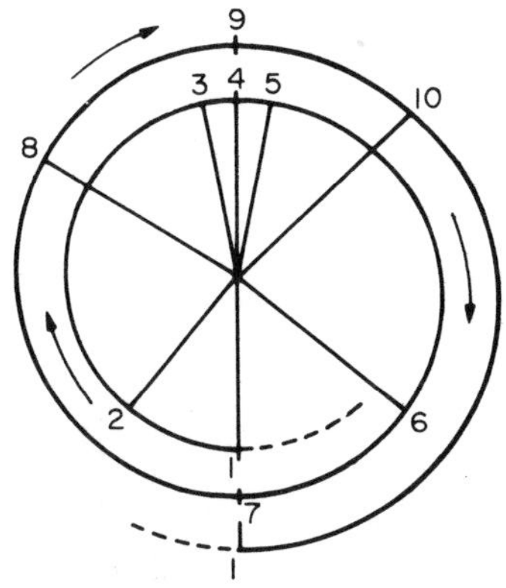


1–2 Completion of aspiration.
2–3 Compression. Air inlet valve has closed, air in cylinder is now compressed to raise its temperature for combustion of fuel.
3–4–5 Fuel injection. Combustion takes place with the rise in pressure.
5–6 Expansion. Combustion completed. Gas pressure forces piston down.
6–7–8 Exhaust. Exhaust valve opened, piston expels exhaust gas on upward stroke.
8–9–10 Overlap. Air inlet valve opened, while exhaust remains open.
10–1 Aspiration. Exhaust valve closed, piston draws air into the cylinder during down stroke.

*Label the diagram with:* **completion of aspiration, compression, fuel injection, expansion, exhaust, overlap, aspiration,** *in the appropriate places.*

# APPLIED TERMINOLOGY

## (A) Terms relating to main engines

*Study this diagram of a two-stroke slow-speed diesel engine and learn the names of the principal parts.*

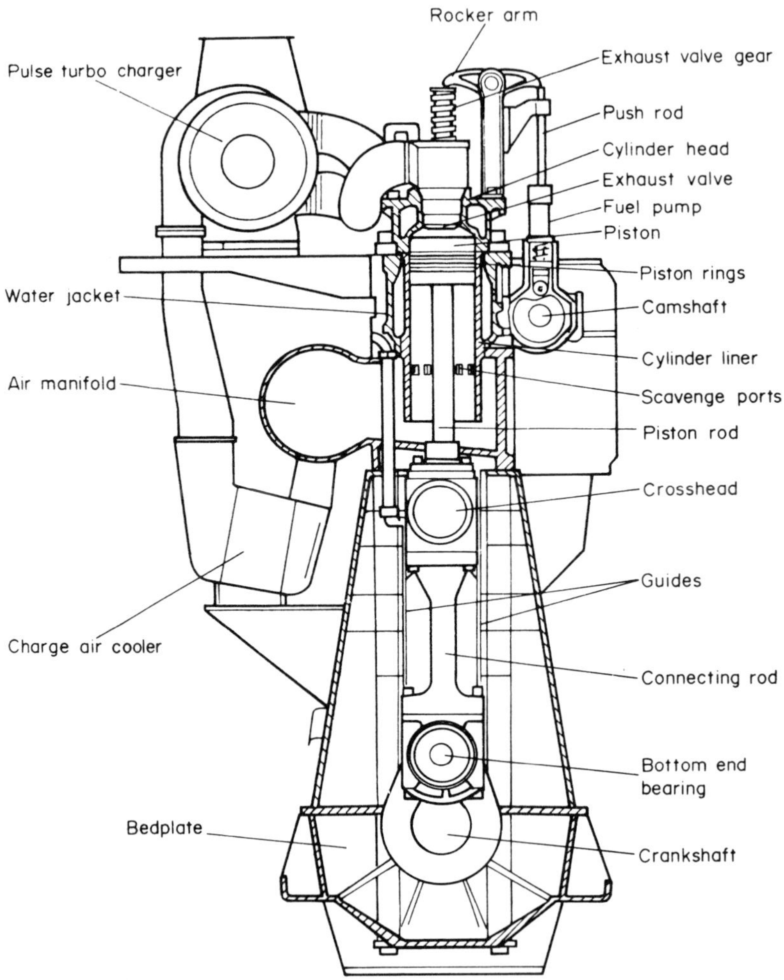

Rocker arm
Exhaust valve gear
Pulse turbo charger
Push rod
Cylinder head
Exhaust valve
Fuel pump
Piston
Piston rings
Water jacket
Camshaft
Cylinder liner
Air manifold
Scavenge ports
Piston rod
Crosshead
Guides
Charge air cooler
Connecting rod
Bottom end bearing
Bedplate
Crankshaft

*Exercise* 1. *Here is a diagram of a four-stroke medium-speed diesel engine. Cover over the diagram above and then see if you can name the parts 1–12 indicated.*

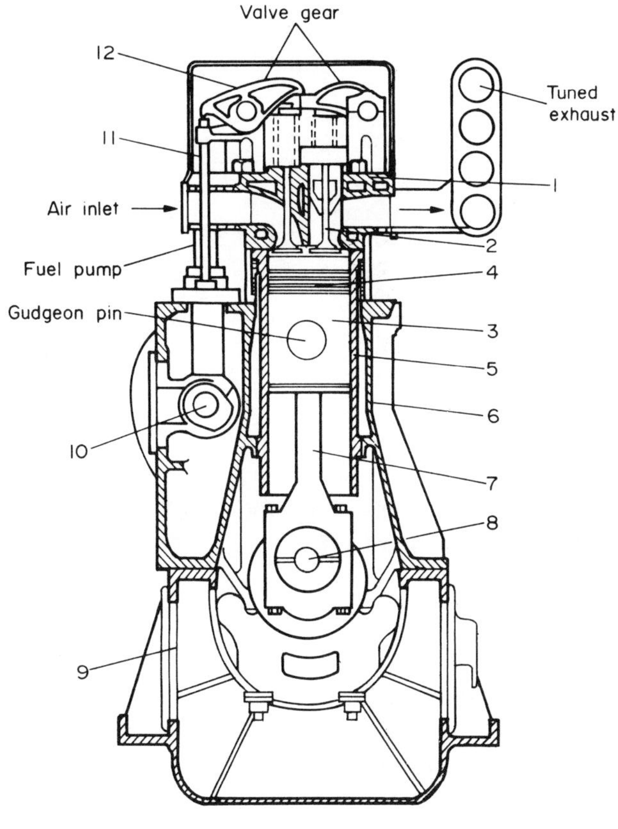


| | | |
|---|---|---|
| 1 ........................ | 5 ........................ | 9 ........................ |
| 2 ........................ | 6 ........................ | 10 ........................ |
| 3 ........................ | 7 ........................ | 11 ........................ |
| 4 ........................ | 8 ........................ | 12 ........................ |

# (B) Measurement

## (viii) SI system

1. Basic units

*Study the table below showing the six basic units of the SI system and their symbols, and the two supplementary symbols. Some have already been used in previous units.*

| Quantity | Unit | Symbol |
|---|---|---|
| length | metre | m |
| mass | kilogramme | kg |
| time | second | s |
| electric current | ampere | A |
| luminous intensity | candela | cd |
| temperature | kelvin | K |
| plane angle | radian | r |
| solid angle | steradian | sr |

2. Derived units I

These are units expressed in terms of base and supplementary units only.

e.g. unit of area = square metre = $m^2$

*Exercise* 1. *Complete the following table:*

(*Note:* Per is expressed by a stroke (/) and indicates that the unit in front of the stroke is divided by the unit after the stroke.)

| Quantity | Unit | Symbol |
|---|---|---|
| area | square metre | ............... |
| volume | cubic metre | ............... |
| velocity | metre per second | ............... |
| angular velocity | radian per second | ............... |
| acceleration | metre per second second | ............... |
| density | kilogramme per cubic metre | ............... |
| momentum | kilogramme metre per second | ............... |

3. Derived units II
Some units have special names. These are shown in the table below.

| | | |
|---|---|---|
| frequency | hertz | Hz |
| force | newton | N |
| work, energy, heat | joule | J |
| power<br>heat-flow rate | watt | W |

4. Derived units III
Some units are expressed in terms of other derived units only or other derived units and base-units.
e.g. unit of stress = newton per square metre = $N/m^2$

*Exercise* 2. *Complete the table below.*

| Quantity | Unit | Symbol |
|---|---|---|
| torque | newton metre | .............. |
| stress | .................................... | $N/m^2$ |
| pressure | newton per square metre | .............. |
| intensity of heat flow rate | .................................... | $W/m^2$ |
| thermal conductivity | watt per metre degree celsius | .............. |
| coefficient of heat transfer | .................................... | $W/m^2\ °C$ |
| heat capacity | joule per degree celsius | .............. |
| specific heat capacity | .................................... | J/kg °C |

# GUIDED WRITING

## (A) Description of an engine governor

*Write a description of how this engine governor works.*

*Stage* 1. Study the diagram of the governor.

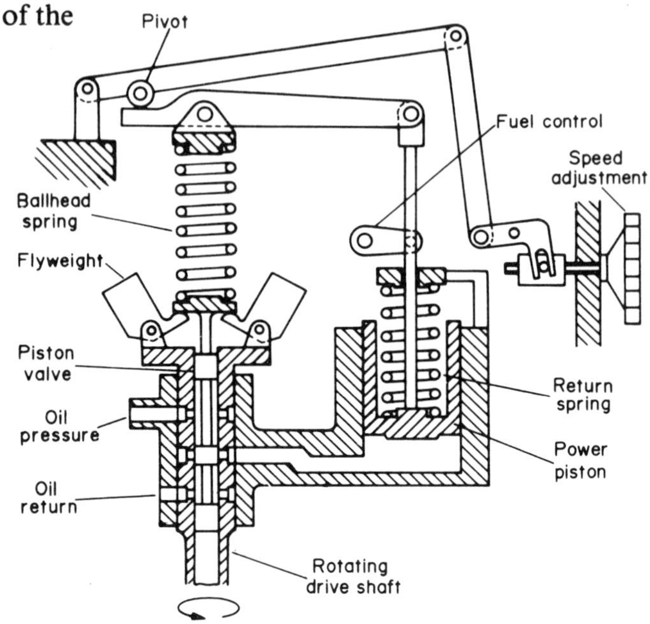


*Stage* 2. *Join some of these sentences together to form an introductory paragraph.*

The diagram above shows a hydraulic engine governor. This governor may be fitted to control the speed of an auxiliary diesel engine. The governor incorporates two systems. The one consists of the mechanical ballhead. This senses any change in the engine speed. The other consists of the hydraulic piston valve. It also consists of the power piston. This operates the fuel pump control setting. Any alteration in speed setting can be made by altering the speed adjustment control.

*Stage* 3. *Now study the flow chart below which shows what happens when the engine speed rises. (Note a single arrow (⟶) indicates that one action **causes** another, a double arrow (⟶⟶) that one action **allows** another to happen.)*

When the engine speed falls below the desired limit . . .

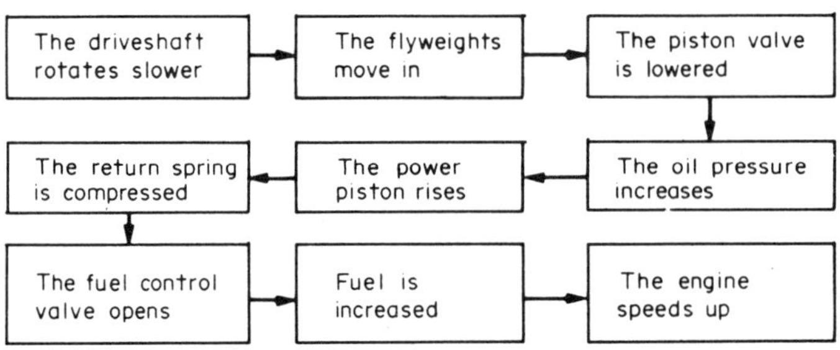

*Now join these together to form a paragraph, using any of the methods of joining cause and effect relations practised so far.*

*Stage* 4. *Complete this flow chart to show what happens when the engine speed rises. It will be the opposite process to the above.*

When the engine speed rises above the desired limit . . .

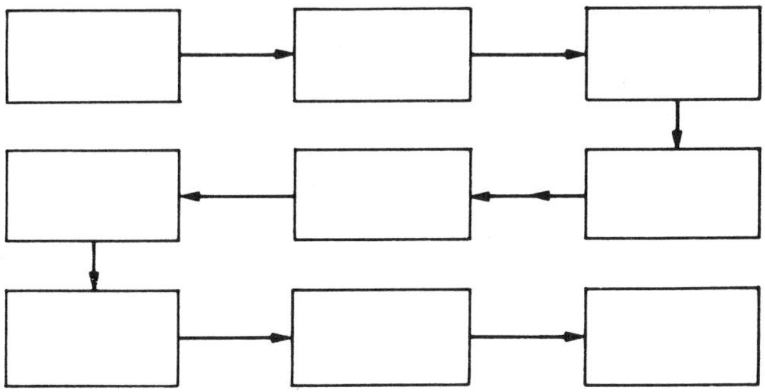

*Now join these together to form another paragraph.*

*Stage* 5. *Put all three paragraphs together to form your description.*

# NOTE-TAKING PRACTICE

## Symbols for intensification or reduction

Here are some more symbols to help you reduce the amount of writing when taking down notes.

1. To intensify or reduce quality and quantity

| | | |
|---|---|---|
| + | : | great, a lot of, a large amount of, e.g. strength + |
| + + | : | very great, a great deal of, e.g. heat + + |
| – | : | a little, a small amount of, e.g. heat – |
| – – | : | very little, a very small amount of, e.g. food – – |

2. To express change

| | | |
|---|---|---|
| ↗ | : | increases, grows, rises, e.g. temperature ↗ |
| ↘ | : | decreases, grows less, falls, lowers, e.g. efficiency ↘ |

**3. To express relations of cause and effect**

⟶ : causes, leads to, results in, produces, gives rise to

⟵ : is caused by, results from, arises from, is due to

In order to take notes we must first understand what the lecture is about. Comprehension can be helped a great deal by the use of a diagram handed out by the lecturer or put up on the blackboard, or by using an overhead projector. These diagrams will help with vocabulary and may be referred to by the lecturer to help explain a system or process.

## **Lecture:** Boilers

*Study the two diagrams below before listening to the lecture on boilers. This will give you some idea of what the words mean and how the systems work.*

*Now listen to the lecture. You will find that a lot of information is already given in the diagrams. Add anything else you think necessary to the diagrams and complete the outline below by adding any other information worth recording, using symbols, abbreviations and other note-taking techniques already practised.*

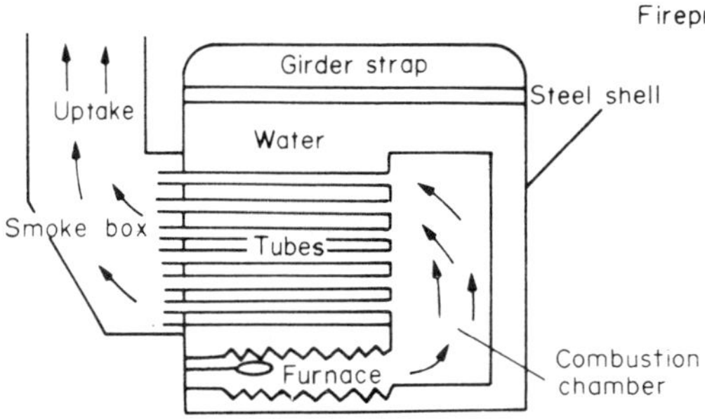


(a) Scotch boiler

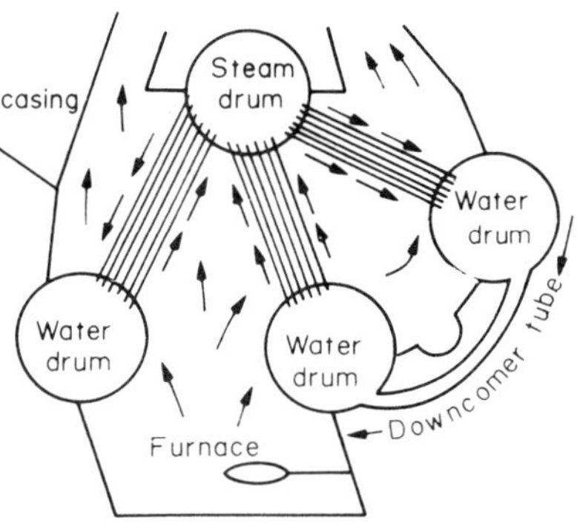


(b) Water-tube boiler

**Title** ..........................................................................................

Two basic types:

(a)..........................................................................................

(b)..........................................................................................

(a) Scotch boilers

..........................................................................................

..........................................................................................

..........................................................................................

..........................................................................................

(b) Water-tube boilers

..........................................................................................

..........................................................................................

..........................................................................................

Boiler mountings include:

safety valves – to release excess steam

main stop valve – ..........................................................................

..................... – ..........................................................................

..................... ..........................................................................

..................... ..........................................................................

..................... ..........................................................................

# GUIDED WRITING

## (B) Boilers

*Write a general description of the two basic types of boiler, using the diagrams and your notes to help you.*

*Your description should be made up of the following paragraphs:*

*Paragraph* 1. Introduction.

*Paragraph* 2. Description of what a Scotch boiler consists of.

*Paragraph* 3. Description of the flow of hot gases through the system.

*Paragraph* 4. Some general characteristics of the Scotch boiler – its uses, efficiency etc.

*Paragraph* 5. Description of what a water-tube boiler consists of.

*Paragraph* 6. Description of the flow of gases through the system.

*Paragraph* 7. Mention some of the boiler mountings fitted to a water-tube boiler.

# UNIT IX

# AUXILIARY MACHINERY

## READING COMPREHENSION

### (A) Functions of auxiliary machinery

We usually have a purpose for reading a passage. If we know nothing about a subject, our purpose may be to gain a general knowledge of it. Later we may want to read about aspects of it in greater detail. Comprehension will be made easier by giving ourselves a definite purpose based on our understanding of what the passage is going to be about from the title and the introductory paragraph. This purpose will help focus our attention.

*Read through this passage and note down* (1) *the different functions that auxiliary machinery is designed to carry out and* (2) *examples of the equipment designed to carry out those functions.*

Besides running and maintaining the main propulsion machinery of the ship, the Engineer Officer has a great deal of auxiliary machinery to look after. Auxiliary machinery covers everything mechanical on board ship except the main engines and boilers. It includes almost all the pipes and fittings and the equipment needed to carry out a number of functions. These functions may be summarized as follows.

To supply the needs of the main engines and boilers. Air compressors are used to supply compressed air for starting engines. Coolers are used for cooling either oil or water. Water for the boilers is also heated before being admitted into the boiler by feed water heaters. This increases the efficiency of the boiler.

To keep the ship dry and trimmed. This is done through the bilge and ballast pumping systems. The former removes water which has gathered in machinery, cargo and other spaces. The latter pumps water into and out of ballast tanks. In general cargo ships, these systems are usually interconnected and served by the same pumps. In tankers and other bulk carriers, these systems are entirely separate, because these ships may need to ballast at 12,000 tonne/hour and therefore need larger pumps.

To supply domestic needs such as fresh water from distillation plant, sanitation from sewage plant and heating and ventilation from heaters and air-conditioners.

To apply the main power of the engines for propulsion and manoeuvring. The engine power is transmitted to the propeller by a line of steel shafting. This is made up of the thrust shaft, intermediate shafts and the propeller shaft. Steering gear is also necessary to operate the rudder for manoeuvring.

To supply the ship with electrical power and lighting. This is done by steam or diesel-powered generators.

To moor the ship and handle cargo. Deck machinery is extensive and varied. It can be divided into anchor-handling machinery – windlass and capstans, mooring machinery – winches and capstans, and cargo-handling machinery – winches and cranes. It also includes cargo oil pumps.

To provide for safety. Firefighting and fire detection equipment, lifeboat engines and launching gear are also included.

Responsibility for auxiliary machinery is often delegated to individual engineer officers, each one taking responsibility for the efficient working of certain items. A lot of equipment is duplicated, so that for example, one generator can be overhauled without cutting off the supply of electricity to the ship. Engineer officers on tankers are also involved in operating the cargo pumping machinery, although the pump rooms themselves are often manned by officers from the Deck Department.

# GRAMMAR

## (A) Change of state verbs

Change from one state to another can be expressed verbally by:

1. Specific change of state verbs
   e.g. to melt, to freeze, to condense, to rot, to rust

2. Verbs formed by adding **-ify**, **-ize (-ise)** to nouns and adjectives
   e.g. solid ⇒ to solidify     vapour ⇒ to vaporize
   liquid ⇒ to liquefy     crystal ⇒ to crystallize
   (*Note:* changes of spelling are often necessary.)

3. General change of state verbs
   **to become + adjective/noun**
   e.g. When air is compressed, it becomes hot.
   When copper and tin are mixed, they become bronze.

**to get + past participle/adjective**
e.g. With poor lubrication, pistons get worn.

**to turn + colour**
e.g. Blue litmus paper turns red, if immersed in acid.

**to change into + noun**
e.g. Water changes into ice when the temperature drops below 0°C.

**to convert/be converted into + noun**
e.g. At an oil refinery crude oil is converted into different oil products.

*Exercise* 1. *Complete the following sentences by using one of the verbs above in the appropriate form:*

(a) An internal combustion engine .............. chemical energy into mechanical energy.
(b) With poor lubrication engine parts ........... worn.

(c) When gas is cooled below a certain temperature, it ..............

(d) The reciprocating motion of a piston .............. a rotary motion by a connecting rod and crank.

(e) Water .............. steam when it reaches boiling point.

(f) Natural fibre ropes .............., if allowed to remain wet.

(g) When red litmus paper is placed in an alkali, it .............. blue.

(h) Liquids .............. less dense, when they are heated.

(i) The sky often .............. red at sunset.

## (B) Comparisons

Comparative statements point out the difference or the equality between things.

Statements pointing out the difference can be made with:

1. **Adjectives** and **adverbs** in the comparative form + **than**
   e.g. Cruisers are larger than frigates.
   Water-tube boilers are more efficient than Scotch boilers.
   Light travels more quickly than sound.

2. **More** or **less/fewer** with **noun + than**

e.g. Less time is needed to load a container ship than a 'tween deck vessel.
There are fewer passenger liners today than there were thirty years ago.
More money is now spent on fuel than before.

(*Note:* **less** is used with uncountable nouns, **fewer** with countables.)

*Exercise* 1. *Rewrite the following sentences so that the part after* **than** *is put first, but the meaning stays the same.*

(a) A nautical mile is longer than a statute mile.
(b) Manila rope is more expensive than sisal rope.
(c) A diesel engine is more economical than a petrol engine.
(d) 100°C is hotter than 100°F.
(e) The Chief Officer is senior to the Second Officer.
(f) There are more members of the Catering Department in a passenger liner than in a cargo vessel.
(g) Mercury freezes at a lower temperature than water.
(h) High carbon steel contains more carbon than mild steel.
(i) A ULCC is larger than a VLCC.
(j) A nuclear submarine is faster than a conventional submarine.

Statements making equal comparison can be made by

3. Using **as . . . as.** The negative is **not so/as . . . as**

e.g. The Deck Department is as important as the Engine Department.
The Atlantic Ocean is not as big as the Pacific Ocean.
A Scotch boiler does not work so efficiently as a water-tube boiler.

*With nouns this structure needs care. Study these examples:*

(singular countable noun)

e.g. Steel is not as good a conductor as copper.

(plural countable noun)

A cargo ship does not have as many lifeboats as a passenger liner.

(uncountable noun)

A plane does not carry as much cargo as a ship.

*Exercise* 2. *Rewrite these sentences using* **not so/as . . . as,** *but without changing the meaning.*

(a) A nautical mile is longer than a statute mile.
(b) Manila rope is more expensive than sisal rope.

(c) A diesel engine is more economical than a petrol engine.
(d) 100°C is hotter than 100°F.
(e) High carbon steel is harder than low carbon steel.
(f) Fewer tankers will be built this year than last year.
(g) Mercury has a lower freezing point than water.
(h) A 'tween deck vessel has more decks than a single deck vessel.
(i) Nowadays British ships carry more tonnage than before.
(j) A nuclear submarine has a higher underwater speed than a conventional submarine.

# READING COMPREHENSION

## (B) Fuel oil system

*Read the following description of the fuel oil system of a typical marine diesel engine and trace the circulation of the heavy oil and the diesel oil on the diagram below. Use a different colour for each:*

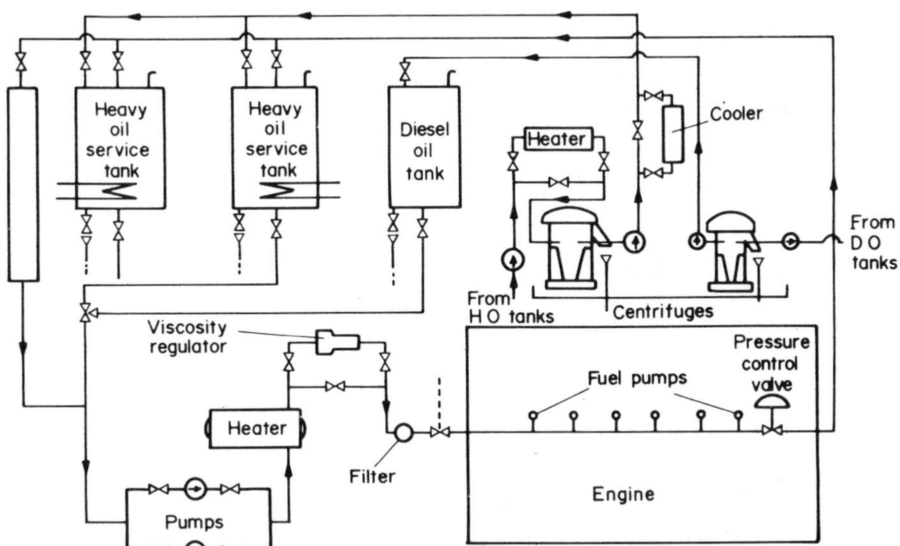


During normal operation on heavy oil at sea, the oil is taken from the HO tanks, where it is stored. First, it is fed through a heater and next through a centrifuge for purification. It may then pass through a cooler, before being discharged to the steam-heated heavy oil service tanks. Two of these are usually fitted and they are used alternately. One tank is in use, while the other is being filled. These tanks are heated to a moderate temperature and self-closing drain valves are fitted to remove any water or sludge which may settle out. The oil is drawn from the service tank in use by pumps and discharged at low pressure to the fuel oil heater. These pumps should be in duplicate. A relief valve on the pumps will return excess pressure to the system. A viscosity regulator is fitted at the heater discharge, through which the oil will pass. This automatically controls the temperature of the oil fuel leaving the heater to maintain its viscosity within close limits. A by-pass must be fitted to the viscosity regulator. The oil is then discharged through a fine strainer to the main engine fuel pump suctions. A pressure control valve is fitted in the system and excess oil returned either to the heavy oil service tanks or to a balancing tank.

A diesel fuel tank is included in the system with its discharge to the primary pump suctions through a change-over valve. By operating this valve the engine may be operated on diesel oil. Change-over should be very gradual to allow temperatures in the system to stabilize. During this period, the excess oil will return either to the heavy oil service tanks or a balancing tank. After it has been pumped from the DO tanks, where it is stored, the diesel oil passes through a centrifuge for purification before entering the diesel oil tank.

# APPLIED TERMINOLOGY

## (A) Terms relating to ancillary services

***Study this diagram of a typical lubricating oil system:***

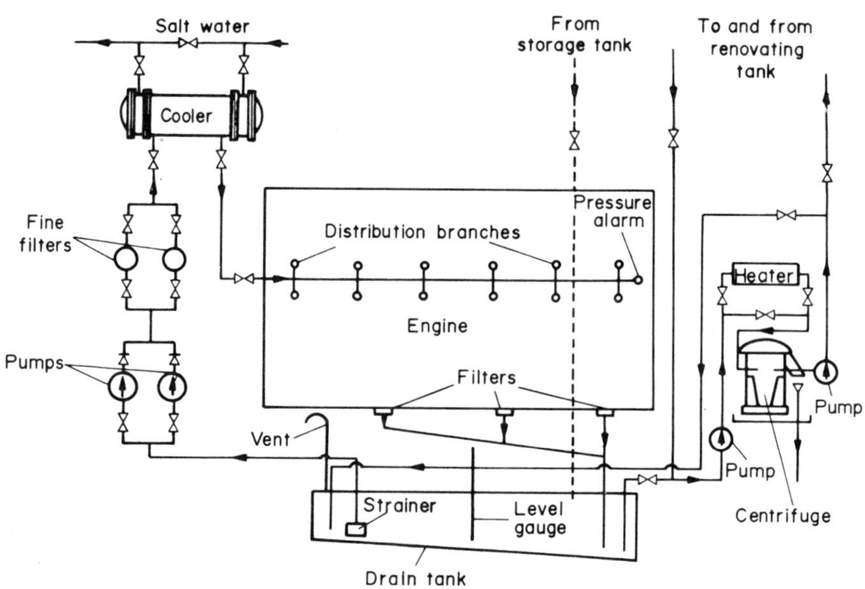

*Exercise* 1. ***Now complete the description below with words from the diagram:***

Pressure pumps draw the oil from the main engine .............. through a suction .............. . The suction strainer should be clear of the lowest point of the tank to avoid intake of any water or sludge. .............. discharge the oil under pressure through fine .............. to the ............ From the cooler the oil passes through the .............. to various parts of the engine. Used oil drains to the crankcase and then through .............. to the drain tank. This is usually in the ship's double bottom. The drain tank should be fitted with an air .............., a .............. and a sounding pipe.

A purifier system is fitted to the oil drain tank. The purifier suction .............. draws oil from the .............. at its lowest point. It may then pass through a .............., before being purified in a .............. After purification, the oil is returned to the .............. at a point next to the suction ..............

## (B) Measurement

### *(ix) SI system (cont.)*

1. Derived units IV – Electrical

*Study the table below showing electrical units in the SI system:*

| Quantity | Unit | Symbol |
|---|---|---|
| electric charge / quantity of electricity | coulomb | C |
| electric potential / electromotive force | volt | V |
| resistance | ohm | Ω |
| capacitance | farad | F |
| self inductance | henry | H |

SI electrical units are interrelated with base units through electrical formulae.

*Exercise* 1. *Write out the following formulae to show the relationship between the symbols. The first is done for you.*

(a) $V = A\Omega$ Volts equal amperes multiplied by ohms.

(b) $C = As$ ..................................................................

(c) $V = \frac{W}{A}$ ..................................................................

(d) $\Omega = \frac{V}{A}$ ..................................................................

(e) $F = \frac{As}{V}$ ..................................................................

(f) $H = \frac{Vs}{A}$ ..................................................................

2. Multiples and submultiples

Decimal multiples and submultiples of SI units are formed by combining the unit with the prefixes listed below.

| Factor by which basic unit is multiplied | Prefix | Symbol |
|---|---|---|
| $10^9$ (1,000,000,000) | giga- | G |
| $10^6$ (1,000,000) | mega- | M |
| $10^3$ (1,000) | kilo- | k |
| $10^{-1}$ (0.1) | deci- | d |
| $10^{-2}$ (0.01) | centi- | c |
| $10^{-3}$ (0.001) | milli- | m |
| $10^{-6}$ (0.000001) | micro- | $\mu$ |

*Exercise* 2. *Complete the following table. The first line has been done for you.*

| | | | |
|---|---|---|---|
| 1,000,000 J | = one megajoule | = 1 MJ | = $10^6$ J |
| 1,000 N | = ..................... | = 1 kN | = $10^3$ N |
| 0.001 V | = ..................... | = ........... | = $10^{-3}$ V |
| 0.000001 Ω | = ..................... | = ........... | = ........... |
| ............... | = one kilojoule | = ........... | = ........... |
| 0.001 A | = ..................... | = ........... | = ........... |

*(Continued overleaf)*

| | | | |
|---|---|---|---|
| ................ | = ...................... | = 1 MW | = ............ |
| ................ | = ...................... | = ............ | = $10^3$ V |
| ................ | = one centimetre | = ............ | = ............ |
| 0.000001 m | = ...................... | = ............ | = ............ |

# GUIDED WRITING

## (A) Description of types of pump

*Write a description of the different types of pump used on board ship.*

*Stage* 1. *Join the following groups of sentences together to form paragraphs. Use any of the linking words you have learned so far.*

Centrifugal pumps consist of an impeller inside a casing.
This impeller rotates at high speed.
The liquid is thrown by centrifugal force against the surrounding casing.
The liquid enters through a suction pipe at the centre.
The liquid is then discharged through the delivery outlet.

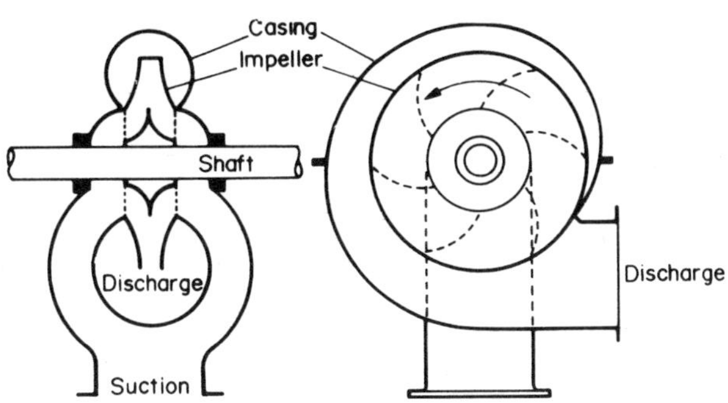


Centrifugal pump

The gear-wheel pump is an example of a rotary displacement pump.
It is used for pumping lubricating and fuel oils.
It consists of interlocking gear wheels.
The gear wheels rotate.
Each tooth on both wheels leaves a vacuum for liquid to flow into.
The next tooth to enter forces the liquid out.
The liquid enters a space, is carried round and then forced out into the delivery tube.

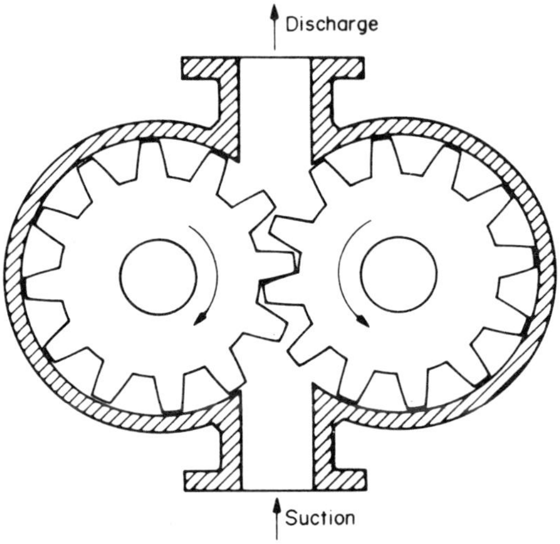


Gear-wheel pump

In displacement pumps the volume of the pump chamber is increased by raising the piston.
This causes a vacuum.
Into this vacuum the liquid is drawn from the suction pipe.
The piston is then lowered.
This decreases the volume of the pump chamber.
It forces the liquid out into the delivery pipe.

A simple kind of reciprocating displacement pump is the single-acting ram pump.
This consists of a ram moving up and down inside a chamber.
The chamber is fitted with a non-return suction valve and a non-return delivery valve.
When the piston moves up, a vacuum is formed in the chamber.
Liquid is drawn into the chamber through the suction valve.
Then the piston moves down.
This creates a force on the liquid.

The suction valve is closed by this force.
The liquid is forced out through the delivery valve.

Another type of reciprocating displacement pump is the double-acting piston pump.
This works on the same principle as the single-acting pump.
The chamber is fitted with suction and return valves at the top and bottom.
The liquid can be drawn in and discharged on each stroke.

Pumps are used on board ship.
These pumps can be divided into two main groups.
The two main groups are displacement pumps and centrifugal pumps.

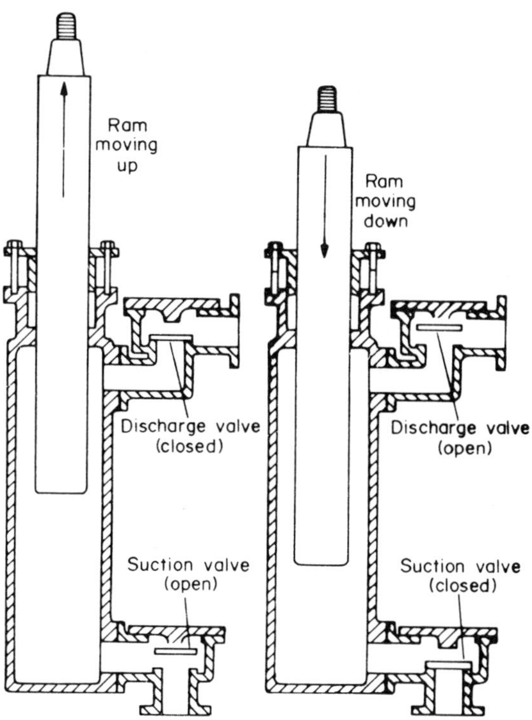


Single-acting ram pump

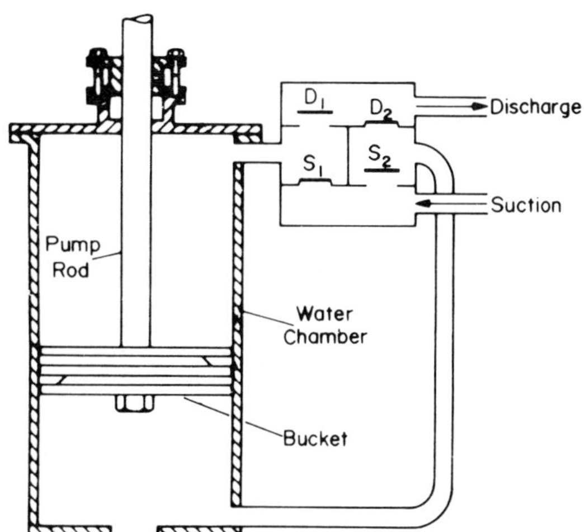


Double-acting piston pump

*Stage* 2. *Use this diagram to sort out your paragraphs into a logical order.*

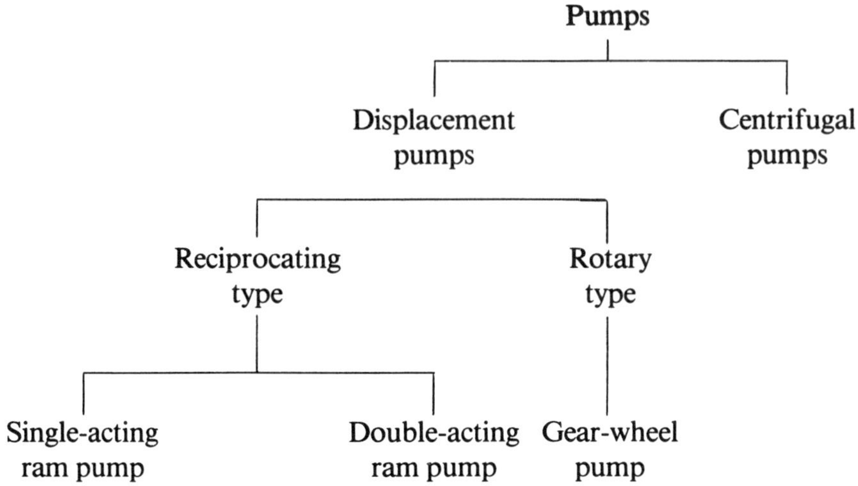


*Stage* 3. *Write out your paragraphs in a logical order to form your description.*

# NOTE-TAKING PRACTICE

## Mathematical symbols

Here are some mathematical symbols used in engineering

| Symbol | Example | Meaning in full |
|---|---|---|
| . | 3.142 | three point one four two |
| + | $a+b$ | *a* plus *b* |
| — | $a—b$ | *a* minus *b* |
| × | speed × distance | speed multiplied by/times distance |
| ÷ | 10 ÷ 5 | ten divided by five |
| / | 20 rev/s | twenty revolutions per second |
| ± | ± 0.25 mm | plus or minus zero point two five millimetres |
| : | 2 : 1 | in the ratio of two to one |
| % | 5% | five per cent |
| $^{2}$ $^{3}$ | $3^2$ $4^3$ | three squared, four cubed |
| $^{6}$ $^{-6}$ | $10^6$ $10^{-6}$ | ten to the power six, ten to the power minus six |
| $\sqrt{}$ | $\sqrt{5}$ $\sqrt[3]{5}$ | the square root of five or root five, the cube root of five |
| $>$ | $x > y$ | *x* is greater than *y* |
| $<$ | $x < y$ | *x* is less than *y* |

Here are some more mathematical symbols used in engineering

| Symbol | Example | Meaning in full |
|---|---|---|
| $=$ | $x=y$ | $x$ equals $y$ |
| $\neq$ | $x \neq y$ | $x$ is not equal to $y$ |
| $\approx$ $\triangleq$ | $x \triangleq y$ | $x$ is approximately equal to $y$ |
| $\propto$ | stress $\propto$ strain | stress is proportional to strain |
| $^\circ$ | $45^\circ$ $18^\circ C$ | forty-five degrees, eighteen degrees Centigrade |
| $'$ | $10^\circ 25'$ | ten degrees twenty five minutes |
| $\infty$ | $1/0 = \infty$ | one over zero equals infinity |

*Practice* 1.

*Now listen to the values read out on the cassette and write them down using the symbols above.*

# GUIDED WRITING

## (B) Description of one piece of auxiliary machinery

*Choose one piece of auxiliary machinery mentioned in the first reading passage of this unit and write a description of what it consists of and how it works.*

*Stage* 1. *Read through the first reading passage again and choose a piece of auxiliary machinery.*

*Stage* 2. *Find out more about this piece of machinery from books in the library and by asking.*

*Stage* 3. *Write a brief introduction saying what the piece of machinery consists of and what it is designed to do. Then explain how it works. Accompany your description with at least one diagram.*

# UNIT X

# MAINTENANCE

## READING COMPREHENSION

### (A) Maintenance on board

Sometimes we read a passage to find specific information. Therefore we look through the passage very quickly until we come to the part which will give us that information. We then read that part more carefully.

*Time yourself to see how long it takes you to find out this information from the following passage on maintenance.*

(a) The basic types of paint used on a ship.
(b) The officer responsible for the overall maintenance of a ship.
(c) The different stages in the preparation of a piece of metal for painting.
(d) The three different coatings which metal can be given to protect it from corrosion.

Much of the work of the Deck Department on board a ship concerns the maintenance of the ship and her fittings. This is the responsibility of the Chief Officer. He and the men in his charge must protect the ship from the damaging effects of salt water, changes in temperature and the action of waves.

The principal material used in building a ship is mild steel, and steel of different types is used for making most fittings and equipment. Unfortunately, steel undergoes a chemical change known as rusting when in contact with air, water or salt solutions. This causes the metal to deteriorate rapidly, unless some form of protection is given.

To try and prevent this corrosion, the metal is coated with cement wash, bitumen and paint. Cement wash is a mixture of cement powder and fresh water. It is used in freshwater tanks and double bottom tanks. Bitumen is used in bilges and peak tanks. It is also used on metal decks before they are sheathed with wood. However, the principal protective coating is paint. There are many types of paint available nowadays in a wide variety of colours and it is no longer necessary for the Boatswain to mix his own. Paints are stowed in the paint locker, which is usually situated under the forecastle head.

The most common kinds of paint found on board ship are as follows: metal primers, which are applied to a bare surface to give protection against rust and to act as a key to the next coat; undercoats, which are used over the primer before the top coat; top coats, which provide a hard-wearing surface and give the required colour; heat-resistant paints for radiators and pipes and for the ship's funnel; non-slip paints for use on weather decks and other suitable surfaces such as companion-ways; and varnishes to give a clear protective coat to woodwork. For painting, the surface of a ship's hull is divided into three distinct areas: the topside, boot-topping and bottom. Topside paint is supplied in the company's colours or in light grey, if the vessel is a warship. It is applied to the area of the ship's hull which is out of the water when the ship is loaded. Boot-topping is applied to that area of the hull which is out of the water when the ship is in the light condition and under water when the ship is loaded. A ship's bottom is given a coating of anti-fouling paint. Anti-fouling paints contain toxicants which are poisonous to marine life. The toxicants have to dissolve out of the paint into the surrounding water in order to be effective.

Before an area can be repainted, proper preparation is essential. The area must be cleaned and washed with a cleaning solution to remove all salt, dirt and oil. The paintwork must then be rinsed with clean fresh water and all trace of the cleaning solution removed. Loose paint and scale is removed with a scraper and any heavy rust with a chipping hammer. Finally, a wire brush should be used on all bare metal to remove the last of the scale, before the first coat of paint is applied.

# GRAMMAR

## (A) Noun compounds

Compounds are common in technical writing. They provide a way of creating new words to describe new objects and processes. They also permit a very condensed form of writing.

Here are three forms of noun compound.

1. **Noun + noun** compounds are formed as follows:

| | | |
|---|---|---|
| A ship for cargo | becomes | a cargo ship |
| A for B | ⇒ | B A |

Many different kinds of relationship are possible in such noun + noun compounds. Some more are listed below:

| | | |
|---|---|---|
| A of B | e.g. | The cover of a cylinder |
| | ⇒ | The cylinder cover |
| A with/has B | e.g. | Water with salt in it |
| | ⇒ | Salt water |
| A contains B | e.g. | The house contains the wheel |
| | ⇒ | The wheelhouse |
| A made of/from B | e.g. | Wire made of copper |
| | ⇒ | Copper wire |
| A in/on/at B | e.g. | The plate at the back |
| | ⇒ | The back plate |
| A operated by B | e.g. | A pump operated by hand |
| | ⇒ | A hand pump |
| A shaped like B | e.g. | A nut shaped like a butterfly |
| | ⇒ | A butterfly nut |
| A uses B | e.g. | A turbine driven by steam |
| | ⇒ | A steam turbine |
| A invented by B | e.g. | Calipers invented by Vernier |
| | ⇒ | Vernier calipers |

*Exercise* 1. *Try and identify the relationships in the following compounds:*

(a) a needle valve
(b) carbon steel
(c) a petrol tank
(d) a wire rope
(e) corner bolts
(f) a foot pump
(g) floor plates
(h) a MacGregor hatch
(i) a diesel engine
(j) a container ship

2. **Present participle + noun** compounds are formed as follows:

valves which regulate ⇒ regulating valves
water which cools ⇒ cooling water

3. **Past participle + noun** compounds are formed as follows:

bolts which have been fitted ⇒ fitted bolts
iron which has been cast ⇒ cast iron

*Exercise* 2. *Form compounds out of the following clauses:*

(a) air which has been compressed
(b) oil which is used for lubricating
(c) ropes which are used for mooring
(d) a joint which has been riveted
(e) a surface which has been machined
(f) a valve which is sticking
(g) oil which has been heated
(h) air which is used for scavenging
(i) a joint which has been welded
(j) rags which are used for cleaning

## (B) Purpose links

Purpose for doing something can be expressed in several ways.

1. By using the **infinitive**
   e.g. He went to college to study engineering.

2. By using **so as** / **in order** } + **infinitive** (this is more formal)
   e.g. He worked hard so as to pass his exams.
   The ship docked in order to take on fuel.
   He ran so as not to be late.

3. By using **so that** + { **may/might, can/could** / **will/would, shall/should** } + **infinitive**
   (This is used when the person to whom the purpose refers is mentioned.)
   e.g. The captain looked through his binoculars so that he could see more clearly.
   He ran so that he would not be late.

*Exercise* 1. *Use your imagination to complete these sentences to show the purpose for the action:*

(a) Ships carry lifeboats ..........................................................

(b) Engines must be regularly serviced ............................................

(c) He opened the window ...........................................................

(d) Safety valves are fitted .......................................................

(e) Ships are taken on sea trials ..............................................

(f) Goggles should be worn ..............................................

(g) He set his alarm clock ..............................................

(h) Warships are painted grey ..............................................

# READING COMPREHENSION

## (B) Fault chart

Information is often presented in the form of a table or chart. When we use a chart, we are usually trying to find a specific piece of information. It is not necessary to read through the whole table to find that information. We skim over the text until something catches our eye and directs us to what we want to know.

*Time yourself to see how long it takes you to find out this information from the following chart showing some common running troubles in a four-stroke medium-speed diesel engine and how they might be remedied.*

(a) What would happen if the valve from the fuel supply tank was not open?
(b) What would you do if the oil viscosity was too high?
(c) Where would you look to see if the starting air pressure was too low?
(d) When is it sometimes necessary to adjust the fuel pump?
(e) What would you do if one of the filters was choked?
(f) What may happen if the fuel pump is incorrectly set?
(g) What would you do if the level of oil in the service tank was too low?
(h) Where would you look if the engine was running too fast?
(i) What would happen if there was water in the fuel?
(j) Where would you look to see if the engine was overloaded?

| Effect | Cause | Where to look | Remedy |
|---|---|---|---|
| Engine refuses to start | (1) No fuel | Supply tank | Fill tank |
| | | Supply tank valve | Open valve |
| | (2) Starting air pressure too low | Starting air pressure gauge | Start air compressor; recharge reservoir |
| | (3) Air in fuel line or pump | Fuel pump | Prime fuel pump; check line for leaks |

| | | | |
|---|---|---|---|
| | (4) Injection nozzles not working | Nozzles | Remove injectors and clean; examine spring |
| | (5) Compression low | Valves | Examine to see if seating properly |
| | | Pistons | Examine piston rings; if supercharged, inlet filter could be dirty |
| | (6) Viscosity of oil too high | Fuel tank | Put on heating steam; circulate fuel lines |
| | (7) Cylinder too cold | Cylinder cooling water pump | Cooling water may need heating |
| | (8) Injection timing wrong | Fuel pump | Adjust |
| Engine starts on air but refuses to pick up firing | (1) Valve open | Inlet exhaust valve | Free, if stuck |
| | (2) Fuel system air locked | Fuel pipes | Prime fuel pumps and test |
| | (3) Filter choked | Fuel filter | Turn on No. 2 filter; clean No. 1 filter |
| | (4) Fuel injector filter choked | Fuel injector | Remove filter; fit spare filter |
| | (5) Fuel pump set incorrectly | Fuel pump timing gear | Reset fuel pump |
| | (6) Level of oil in service tank too low | Service tank gauge | Refill service tank and prime fuel pumps |
| Engine slows down or stops | (1) Governor gear defective | Governor and linkage | Check setting of governor |
| | (2) Fuel injector delivering too little fuel | Fuel pump<br>Injector | Adjust<br>Put in new injector |
| | (3) Water in fuel oil | Filter | Turn on No. 2 filter; clean No. 1 filter |
| | (4) Overloading | Gauges | Look at all gauges; reduce load |
| | (5) Unequal load being developed in one cylinder | Exhaust temperatures | Adjust fuel supply to low cylinder; check injector nozzle and fuel pump delivery valve |
| | (6) Stoppage of cooling water | Circulating pump | Regrind pump gland; reset heat exchanger thermostat |

| | | | |
|---|---|---|---|
| Engine runs fast | (1) Governor gear out of order | Governor and governor links | Clean oil and reset; check connections |
| | (2) Fuel pump not responding to governor | Fuel pump | Examine fuel pump to governor joint |
| Engine works irregularly | (1) Governor gear out of order | Governor and linkage | Examine, clean and oil |
| | (2) Water in fuel | Fuel filter | Change over filter |
| | (3) Overloading | All gauges | Adjust where necessary |
| | (4) Fuel pump valve leaking | Fuel pump | Delivery valve may be stuck open |
| | (5) Fuel pump sticking | Fuel pump | Check spring; clean pump plunger |
| | (6) Fuel delivery differs | Injectors | Adjust until all inject at same pressure |

# APPLIED TERMINOLOGY

## (A) Terms relating to maintenance

*Exercise* 1. *Explain the names of the cleaning tools illustrated below by expanding the compounds:*

e.g. A chipping hammer = a hammer used for chipping

(a) Munday hammer

(b) Straight scraper

(c) Goose-neck scraper

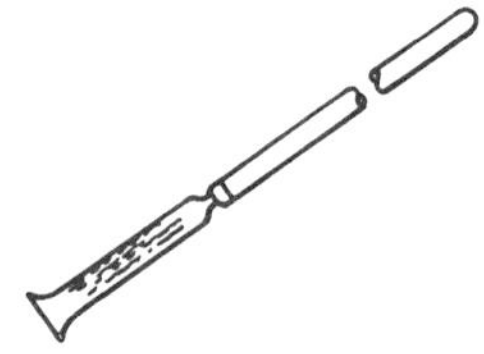

(d) Long-handled scraper

(e) Wire brush

(f) Three-cornered wood scraper

(g) Deck broom

(h) Turks head

*Exercise* 2. *Now do the same with these names of paint brushes:*

(a) Flat brush

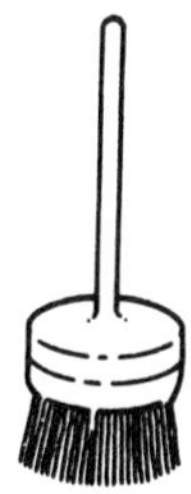

(b) Oval brush

(c) Tar brush

(d) Lining brush

(e) Stencil brush

(f) Pencil brush

## (B) Measurement

### *(x) Number compounds*

1. The length, height, weight etc., of an object can be described in the following way:

e.g. The mast is twenty feet high ⇒it is a twenty-foot mast
The load weighs ten tons ⇒it is a ten-ton load
The size of the gun is six inches ⇒it is a six-inch gun

(*Note:* (a) The scale of measurement is always in the singular.
(b) When the number is written adjectivally, there is a hyphen(-))

2. Number compounds can also be formed by using these prefixes:

| Prefix | Meaning | Examples |
|---|---|---|
| uni- / mono- | one, single | unicolour, monotone |
| bi- | two | biplane |
| tri- | three | tripod |
| quadr- | four | quadrilateral |
| semi- / hemi- | half | semi-skilled, hemisphere |
| multi- / poly- | many | multi-deck, polygon |
| twin- | two the same | twin-masted |

*Exercise* 1. *Explain these compounds:*

(a) a six-inch ruler
(b) a twin-screw vessel
(c) a two-inch nail
(d) a bipod
(e) three-stranded hemp line
(f) multi-grade oil
(g) 15-ton derrick
(h) a uni-flow scavenging air system
(i) 120-fathom coils of rope

# GUIDED WRITING

## (A) Description of causes of cylinder liner wear

In a composition, paragraphs are put together in a logical order. They may be arranged to show a sequence of events, or to support an argument, or to give an explanation. Their order may be based on a classification diagram.

*Read through these three paragraphs about wear in cylinder liners:*

Frictional wear depends on the materials used in the liner and in the piston rings. It takes place between the sliding surfaces of the cylinder liner and the piston rings. It may be caused by inefficient lubrication of the cylinder or by overloading the engine. In addition, engine operating conditions also affect frictional wear. For example, frictional wear may increase if the air and fuel are contaminated.

Corrosion occurs when heavy fuels with a high sulphur content are burned. It is caused by acids which are formed during combustion. Sulphuric acid corrosion may be caused if the cooling water temperature is too low. This allows moisture to condense in the cylinder cooling jacket and consequently sulphuric acid may form in the cylinder.

Abrasion is caused by hard particles. These may be the product of combustion or mechanical wear. Hard particles may also be produced by corrosion.

The relationship between paragraphs can be made clearer by **linking paragraphs** and **bridging sentences.**

There are three basic types: **introductory, transition** and **summary.**

1. Here is an example of an **introductory paragraph**:
   There are a number of reasons for wear in cylinder liners. Wear may be due to friction, corrosion and abrasion. Each of these may have a number of causes.

2. Here is an example of a **bridging sentence**:
   Having discussed the causes of frictional wear, let us go on to consider the reasons for corrosion.
   (These are put at the beginning of the paragraph they introduce.)

3. Here is an example of a **summary paragraph**:
   We have shown, then, that cylinder liner wear is caused by friction between the liner and the piston rings, by corrosion – mainly from burning heavy fuels – and by abrasion from the products of wear corrosion and combustion.

*Now put together the three paragraphs on the causes of wear in cylinder liners, using the paragraphs and sentences above. You will need to write one other linking paragraph or bridging sentence.*

# NOTE-TAKING PRACTICE

## **Lecture:** Maintenance schedule of marine diesel engine

*Before listening to the lecture in this section, study any diagram showing the principal parts of a marine diesel engine. (There is an example on page 122). This will help you with vocabulary used in the lecture.*

*Now listen to the lecture, which will give you a general idea of the maintenance schedule of a marine diesel engine. Make brief notes on the outline table below. The first notes have been done for you.*

| Interval | Engine part | Maintenance |
|---|---|---|
| frequently | fuel pumps<br>pistons | exam & adjust if nec<br>exam for cracks |
| 6 wks | fuel valves<br>atomzrs & fltre | |

# GUIDED WRITING

## (B) Description of the maintenance programme for a marine diesel engine

*Write a description of the maintenance programme for a particular make of marine engine.*

*Give a brief introduction, then divide the maintenance programme up into suitable periods of time. Write a paragraph on each describing the maintenance which has to be done.*
(*Note:* In order to write this description you will need to consult a good book on marine engines and/or a maker's servicing manual.)

# UNIT XI

# SAFETY ABOARD

## READING COMPREHENSION

### (A) Collision Regulations 1977 (extracts)

*Read through the following extracts, which have been taken from the Collision Regulations 1977 as laid down by The Intergovernmental Maritime Consultative Organization (IMCO).*

The Regulations are in four parts embodying thirty-eight Rules and four annexes. Part (B) covers steering and sailing rules.

**Section II. Conduct of vessels in sight of one another**

RULE 11

*Application*

Rules in this Section apply to vessels in sight of one another.

RULE 13

*Overtaking*

(*a*) Notwithstanding anything contained in the Rules of this Section any vessel overtaking any other shall keep out of the way of the vessel being overtaken.

(*b*) A vessel shall be deemed to be overtaking when coming up with another vessel from a direction more than 22.5 degrees abaft her beam, that is, in such a position with reference to the vessel she is overtaking, that at night she would be able to see only the sternlight of that vessel but neither of her sidelights.

(*c*) When a vessel is in any doubt as to whether she is overtaking another, she shall assume that this is the case and act accordingly.

(*d*) Any subsequent alteration of the bearing between the two vessels shall not make the overtaking vessel a crossing vessel within the meaning of these Rules or relieve her of the duty of keeping clear of the overtaken vessel until she is finally past and clear.

## Rule 14

### *Head-on situation*

(*a*) When two power-driven vessels are meeting on reciprocal or nearly reciprocal courses so as to involve risk of collision each shall alter her course to starboard so that each shall pass on the port side of the other.

(*b*) Such a situation shall be deemed to exist when a vessel sees the other ahead or nearly ahead and by night she could see the masthead lights of the other in a line or nearly in a line and/or both sidelights and by day she observes the corresponding aspect of the other vessel.

(*c*) When a vessel is in any doubt as to whether such a situation exists she shall assume that it does exist and act accordingly.

## Rule 15

### *Crossing situation*

When two power-driven vessels are crossing so as to involve risk of collision, the vessel which has the other on her own starboard side shall keep out of the way and shall, if the circumstances of the case admit, avoid crossing ahead of the other vessel.

## Rule 16

### *Action by give-way vessel*

Every vessel which is directed to keep out of the way of another vessel shall, so far as possible, take early and substantial action to keep well clear.

## Rule 17

### *Action by stand-on vessel*

(*a*) (i) Where one of two vessels is to keep out of the way the other shall keep her course and speed.

(ii) The latter vessel may however take action to avoid collision by her manoeuvre alone, as soon as it becomes apparent to her that the vessel required to keep out of the way is not taking appropriate action in compliance with these Rules.

(*b*) When, from any cause, the vessel required to keep her course and speed finds herself so close that collision cannot be avoided by the action of the give-way vessel alone, she shall take such action as will best aid to avoid collision.

(*c*) A power-driven vessel which takes action in a crossing situation in accordance with sub-paragraph (*a*)(ii) of this Rule to avoid collision with another power-driven vessel shall, if the circumstances of the case admit, not alter course to port for a vessel on her own port side.

(*d*) This Rule does not relieve the give-way vessel of her obligation to keep out of the way.

### RULE 18

*Responsibilities between vessels*

Except where Rules 9, 10 and 13 otherwise require:

(*a*) A power-driven vessel underway shall keep out of the way of:

(i) a vessel not under command;
(ii) a vessel restricted in her ability to manoeuvre;
(iii) a vessel engaged in fishing;
(iv) a sailing vessel.

. . . . . . . . .

*Exercise* 1. *From your understanding of the rules above, what should the black vessel do in each situation?*

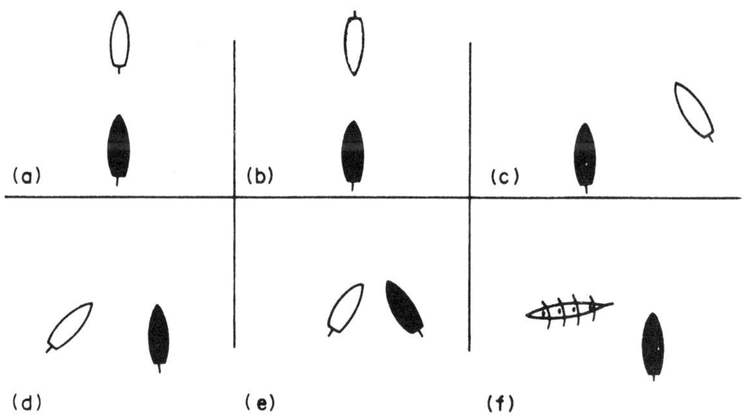


# GRAMMAR

## (A) Obligation

Commands in English can be expressed in a number of ways:

1. By using the **imperative.** This has the same form as the infinitive without *to*

e.g. Be careful! Close the door!
Don't smoke! Don't make a noise!

2. By using **must/must not**
   e.g. You must be careful! Doors must be kept closed!
   Hatches must not be left uncovered!

3. By using **be to + infinitive.** This is an indirect form of command often used in official notices
   e.g. You are to answer all the questions.
   Officers are to be back on board by 1800.
   Lights are not to be left on unnecessarily.

4. By using **will.** This is used to express a severe command
   e.g. Regulations will be observed at all times.
   You will not be allowed ashore until further notice.

Mild commands in the form of advice and recommendations can be made as follows:

5. By using **should/ought to**
   e.g. You should read all safety regulations.
   The work ought to be finished by 1630.
   Tools should not be left lying about.

Strong advice and recommending the wisest course of action can be made as follows:

6. By using **had better**
   e.g. You had better report sick.
   You had better not touch that electric wire.

*Exercise* 1. *Read the following advice based on an extract from a safety handbook for engineer officers. Using the imperative, make up a notice summarizing the main points that might be put on the door of an unmanned machinery space. The first is done for you:*

A seaman should not enter an unmanned machinery space unless he has been given permission by the officer in charge. While in the space, reports by telephone must be made at regular intervals to the duty Deck Officer. A seaman should only do the task which he has been specifically told to do. If any job is beyond his unaided capability, he should obtain assistance. The Engineer Officer in charge must be informed by the seaman in person when he leaves the space.
The maintenance of adequate lighting and clean conditions is essential at all times.
Alarm circuits should be tested regularly to the manufacturers' schedules.
At all times, personnel working in unmanned machinery spaces should be on their guard against the sudden starting of automated machinery.

1. Do not enter without permission from the officer in charge.

2. Report by telephone .......................................................

3. ..........................................................................

*Now continue. There are about nine orders in all.*

*Exercise* 2. *Use your imagination to respond to these statements. (Use* ***should/ought to*** *for advice, use* ***had better*** *for making a stronger recommendation).*

(a) 'I'm spending my leave in Hong Kong this year.'
(b) 'The temperature level is far too low.'
(c) 'I think he's stopped breathing.'
(d) 'I overslept again this morning.'
(e) 'One of the cadets has a terrible cough and sore throat.'
(f) 'I don't know how this paint spray gun works.'
(g) 'I've just spilled a can of oil over the floor.'
(h) 'The exam starts in two minutes time.'
(i) 'My eyesight seems to be getting worse.'
(j) 'The weather's turning bad.'

## (B) Conditionals

There are three basic types of conditional sentence, which are sometimes referred to as **real, unreal** and **impossible** conditionals. Each type is illustrated below:

1. **Real** conditions are those which may or may not be fulfilled.
   **Pattern: if + present (simple, continuous, perfect), future tense/modal/imperative**
   e.g. If I pass my exams, I shall celebrate.
   If you are working on board, you should wear the right clothing.
   If you hear the alarm bells ring, muster on deck immediately.

2. **Unreal** conditions are those which are not expected to be fulfilled.
   **Pattern: if + past (simple, continuous), conditional**
   e.g. If the ship was sinking, we would launch the lifeboats.

3. **Impossible** conditions are those which are impossible because they are contrary to past fact.
   **Pattern: if + past perfect (simple, continuous), conditional perfect**
   e.g. If the cargo had arrived, it would have been loaded.

(*Note:* (a) Types 1 and 2 refer to present or future time.
Type 3 to a past situation.
(b) The conditional (**if**) clause can be put second in all three types.)

*Exercise* 1. *Find the correct sentences in the right hand column to match the sentences on the left. Then write them out.*

| | |
|---|---|
| (a) If the weather is bad, | I would recognize him. |
| (b) If we had a torch, | you will be electrocuted. |
| (c) If I had passed the exam, | we will work below deck. |
| (d) If you touch that wire, | I would have bought it. |
| (e) If I saw him again, | we would be able to see. |
| (f) If the car had been cheaper, | I would have been promoted. |

*Exercise* 2. *Use your imagination to complete these sentences:*

(a) If I fell overboard ..............................................................

(b) ......................................................, he would have been saved.

(c) ................................................, if we finish painting the funnel.

(d) If the radio had been manned ..................................................

(e) ....................................., you would report it to the duty officer.

(f) They will miss the boat, ..........................................................

# READING COMPREHENSION

## (B) Code of safe working practices (extracts)

Read the following extract taken from the *Code of Safe Working Practices for the Safety of Merchant Seamen, Officers, engineers:*

### *Working clothes and personal protective equipment in machinery spaces*

**1.** When working clothes are selected for use in machinery spaces, consideration should be given to the hazards to which the wearer may be exposed. Personnel should ensure that clothes fit well and have no loose flaps or strings. External pockets, if any, should be as few and as small as possible. A shirt with short sleeves should be worn in preference to one with rolled up sleeves.

**2.** Loose, torn or ragged garments are a hazard when working near moving machinery. Special attention is drawn to the hazards of neck ties, sweat rags slung round the neck, and the wearing of finger rings in the machinery spaces.

**3.** Special caution is necessary when working near machinery in motion, particularly in close proximity to a revolving shaft.

**4.** Suitable goggles should always be worn when handling chemicals, grinding, drilling, working a lathe, scaling, hammering, using a cold chisel or doing any work of a similar nature.

**5.** Every effort should be made to ensure that any oil falling on floor plates be removed as quickly as possible. Nevertheless, floor plates tend to become slippery and this, combined with the movement of the ship at sea, makes footholds insecure. When floor plates are slippery with oil, there is a risk that heavy machinery which is being lifted may be inadvertently dropped, causing severe foot injuries. The risk of injury is reduced by the use of industrial footwear (not having steel studs and preferably with soles having oil-resistant anti-slip characteristics).

**6.** Where overhaul of engine room machinery creates special hazards, safety helmets should be worn.

**7.** Personnel required to work in machinery spaces which have high noise levels should be provided with suitable ear defenders, which may be ear plugs or ear muffs. Where noise is intense a combination of both may be desirable.

**8.** Clothing should not be put to dry in any machinery space.

*Exercise* 1. Complete the table below to summarize the information given in the extract. Part of it has been completed for you.

| Part of the body | Type of protection | Situations |
|---|---|---|
| body | well-fitting clothes | working in machinery spaces |
| neck | no ties, no sweat rags | ........................................ |
| fingers | .................................. | ........................................ |
| eyes | .................................. | ........................................ |
| feet | .................................. | ........................................ |
| head | .................................. | ........................................ |
| ears | .................................. | ........................................ |

*Exercise* 2. *Now use the information in the table to write out statements like this:*

If you are working in machinery spaces, you should wear well-fitting clothes in order to protect your body.

If you are working near machinery in motion, you should not wear ties or sweat rags, in order to protect your neck.

# APPLIED TERMINOLOGY

## (A) Terms relating to life-saving appliances

*Exercise* 1. *Try and name the life-saving appliances illustrated below from this list: life-raft, life-jacket, lifeboat, lifebuoy.*

(a) ......................

(b) ....................

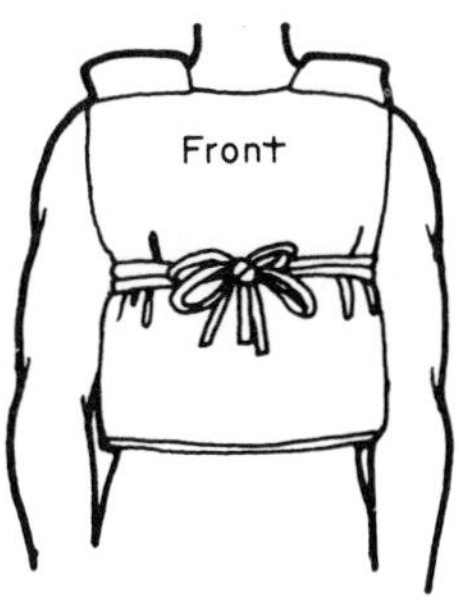


(c) ......................

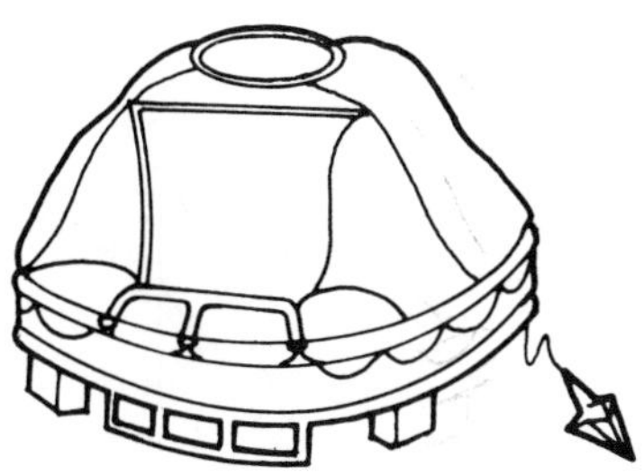

(d) ....................

*Exercise* 2. *State which of the above were to be found on the ship you served in, how many there were of each, and where they were stowed.*

# Terms relating to lifeboats

***Study the illustration showing some of the parts of a lifeboat.***

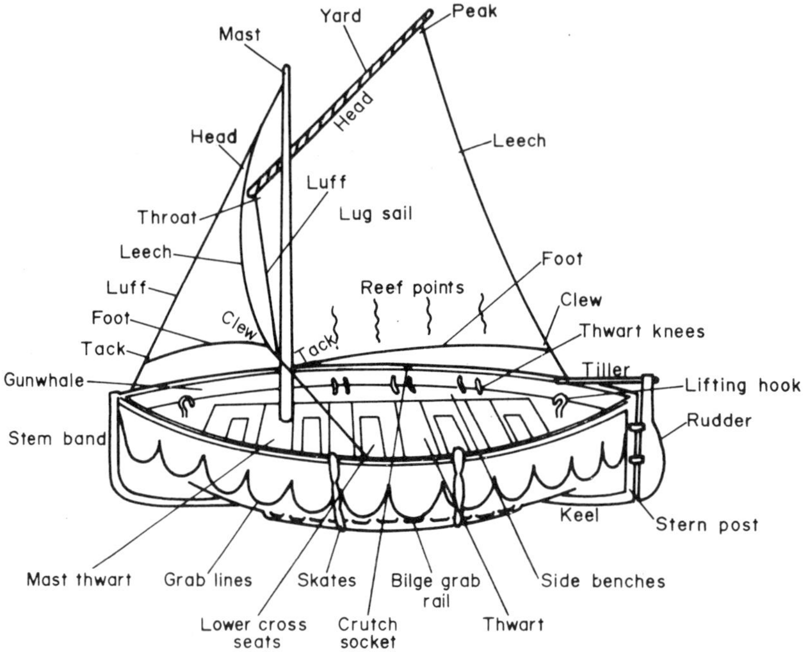


*Exercise* 3. ***Cover the illustration above and see how many terms you can remember by labelling the diagram below.***

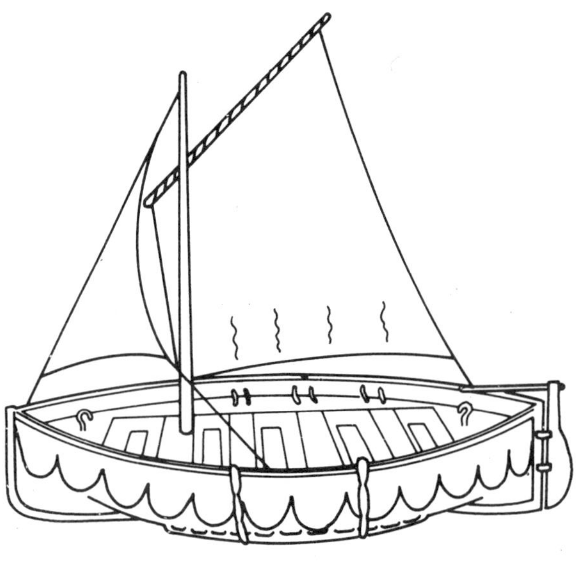

# (B) Measurement

## *(xi) Rope*

Rope is measured by its diameter (*D*) in millimetres (mm). The larger sizes are generally supplied in coils of 220 metres or 120 fathoms (1 fathom = 1.8288 metres).

### The breaking strain of ropes

This can be calculated from the formulae below:

| Material | Size range | Factor |
|---|---|---|
| Grade 1 Manila | 7 mm to 144 mm | $2D^2/300$ tonnes |
| Polythene | 4 mm to 72 mm | $3D^2/300$ tonnes |
| Polypropylene | 7 mm to 80 mm | $3D^2/300$ tonnes |
| Polyester (Terylene) | 4 mm to 96 mm | $4D^2/300$ tonnes |
| Polyamide (Nylon) | 4 mm to 96 mm | $5D^2/300$ tonnes |
| Flexible steel wire rope | | |
| 12 wires per strand | 4 mm to 48 mm | $15D^2/500$ tonnes |
| 24 wires per strand | 8 mm to 56 mm | $20D^2/500$ tonnes |
| 37 wires per strand | 8 mm to 56 mm | $21D^2/500$ tonnes |

### The safe working load

This can be calculated by multiplying by a factor of one sixth.

e.g. The SWL of 20 mm polypropylene rope =

$$\frac{3 \times 20 \times 20}{300 \times 6} = \frac{1200}{1800} = \frac{2}{3} \text{ tonnes}$$

*Exercise* 1. *Calculate the SWL for the following ropes:*

(a) 10 mm grade 1 Manila

(b) 10 mm polythene

(c) 12 mm polypropylene

(d) 15 mm polypropylene

(e) 20 mm polyester

(f) 8 mm polyamide

(g) 14 mm 12 wires/strand

(h) 15 mm 24 wires/strand

# GUIDED WRITING

## (A) Description of lowering a lifeboat

*Write a description of lowering a lifeboat from davits.*

*Stage* 1. *Study the diagram and read the notes below.*

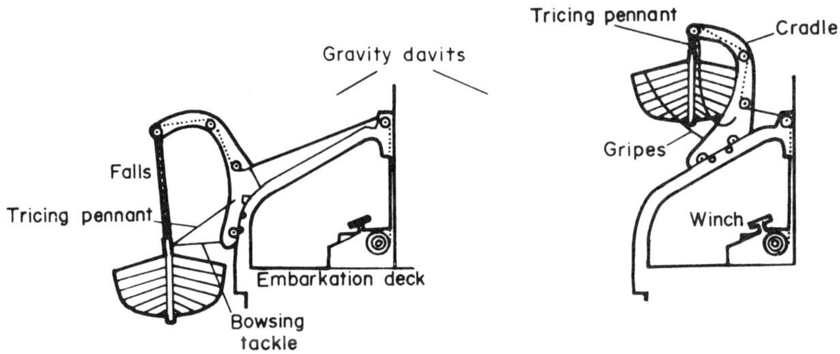


Gravity davits

The coxswain orders the boat's crew to fall in and number off.
The coxswain details members of the crew to do particular jobs.
The harbour pins are cleared.
Two crew members enter the boat.
The plug is shipped.
The toggle painter is secured to the ship.
The gripes are cleared away.
The boat is lowered to the embarkation deck.
The bowsing tackles are secured and the boat bowsed in.
The tricing pennants are slipped.
The passengers and the crew embark.
The bowsing tackles are released.
The boat is lowered to the water.

*Stage* 2. *Write your description adding in the following information.*

(a) The harbour pins prevent the boat being accidentally lowered.
(b) The plug is made of cork.
(c) The plug fits in a hole near the fore-and-aft line of the boat.
(d) The toggle painter prevents the boat from drifting astern, when lowered.
(e) The gripes secure the boat, when in the stowed position.

(f) The bowsing tackles take the weight of the boat off the tricing pennants.
(g) The tricing pennants bring the boat towards the embarkation deck as it is lowered.

# NOTE-TAKING PRACTICE

## Ticking items on a list

It is frequently necessary to check items off from a list to make sure that they are present. This does not entail taking notes but putting a mark, usually a tick (✓), next to the items.

*Study this list of the equipment which should be in every lifeboat. If you do not know what some of the words mean, find out by asking or using a good dictionary.*

5 Oars
6 Crutches
2 Boat hooks
1 Bailer
2 Buckets
2 Plugs
1 Rudder and tiller
Double lifeline round boat
1 Locker
2 Hatchets
1 Lamp
1 Water-tight box
1 Mast
1 Compass
1 Sea anchor
1 Exposure cover

2 Painters
1 Gallon of storm oil
6 Hand flares
2 Smoke signals
4 Parachute signals
1 First-aid kit
1 Waterproof torch
1 Signalling mirror
1 Jack-knife
2 Buoyant heaving lines
1 Manual pump
1 Whistle
1 Fishing line (6 hooks)
1 Copy of the Department of Trade Rescue Table
Food and water

................................ ................................

................................ ................................

*Practice* 1

*Now listen to the list of items being read out. Tick them off on the list above as they are called. If any items are called which are not on your list, add them to the end of the list.*

# GUIDED WRITING

## (B) Answering questions on fires

*Using your knowledge of firefighting and relevant information from other sources (e.g. Chapter 8 in* Seamanship Notes *by Kemp & Young (Stanford Maritime)) try and answer* one *of the following questions:*

(a) Discuss the advantages and disadvantages of the different methods of putting out fires.
(b) Describe the different methods of dealing with fires in machinery spaces.
(c) Describe the different methods of dealing with fires in cargo spaces.

# UNIT XII

# COMMUNICATIONS AT SEA

## READING COMPREHENSION

### (A) List of reading techniques

During this course you have practised a number of techniques to help you with the comprehension of written English. These techniques may be summarized as follows.

When approaching a written text:

1. First look at the title of the chapter or the section heading, if there is one, to see what the passage is about.
2. Read the first paragraph, it may be an introductory paragraph, giving you more detail of what you are about to read.
3. When you have this information, bring to the forefront of your mind any knowledge you already have on the subject as this will help you with vocabulary and general comprehension.
4. Look carefully at any diagrams relevant to the text.
5. If you have not already got a purpose for reading the text, give yourself one based on the introduction. This will help focus your attention.
6. Read the passage through quickly for general comprehension and overall structure. Don't stop for very long on words you don't understand.
7. Read the passage through again more carefully, concentrating on those passages relevant to your purpose. Study the relationships between words and sentences and how paragraphs relate. Try and guess the meaning of words from the context. Use any diagrams to help you with vocabulary and overall comprehension.
8. Test your comprehension of the passage by trying to summarize the information in notes or diagrammatic form. Relate what you have got out of the passage to your purpose for reading. If you have not got the information you wanted, or if you have not understood what the passage is saying, and this is important to you, read it again.

The important things are: (1) to have some idea of what a passage is going to be about before you start; (2) to use any knowledge you already have on the subject and any pictures and diagrams to help you with comprehension; (3) to give yourself a realistic purpose for reading; (4) to check at the end to see if you have achieved your purpose.

# GRAMMAR

## (A) Modal expressions of certainty

*Study this diagram showing degrees of certainty expressed with modal verbs.*

| | | | |
|---|---|---|---|
| | **will/will not** | | 100% |
| | **must/cannot** | | ‖ |
| If you do *x*, *y* | **may/may not** | happen | 50% |
| | **might/might not** | | 40% |
| | **could** | | ‖ |
| | | | 0% |

(*Note:* (1) The past is formed by adding **have**, *e.g.* **must have, may not have.**
(2) **Will** expresses a fact, **must** expresses a conclusion.)

*Exercise* 1. *Fill in the blanks using any of the words above in the present or past tense as appropriate.*

(a) If fuel oil is dirty, it .............. choke the fuel valves.

(b) A naval officer with four gold cuff bands on his sleeve .............. have the rank of Captain.

(c) If clothing with loose flaps is worn in machinery spaces, it ............. get caught up in the moving parts of the machinery.

(d) Distress signals must always be repeated, because the first one ......... been heard or seen.

(e) If a crew member touches a live wire carrying 440 V AC, he ........... be electrocuted.

(f) A lifeboat whistle and jack-knife ........... get lost, if a lanyard is not attached to them.

(g) During a semaphore message, if the signaller makes a series of 'E' signals, he ........... made a mistake.

(h) Always follow the instructions carefully when using pyrotechnics, then you ........... get hurt.

Certainty can also be expressed by using these phrases:

| | |
|---|---|
| is/are certain (not) to | is/are (un)likely to |
| is/are sure to | will probably |
| will perhaps | may possibly |

Where would you place these on the certainty scale?

*Exercise* 2. *Replace* ***may, might*** *etc., in the exercise above with these phrases.*

*Exercise* 3. *Use your imagination to finish these sentences. Use any of the words and phrases expressing degrees of certainty that you have learnt.*

(a) If navigation lights are not displayed at night..............................

(b) If lighted cigarettes are thrown down ........................................

(c) If a seaman falls into a cargo hold ...........................................

(d) If cargo is not properly secured ...............................................

(e) If you do not wear goggles when using a lathe ..........................

(f) If ropes are not looked after ..................................................

## (B) Non-finite clauses

During this course, we have looked at a number of ways of linking statements. We have also seen that some of these can be reduced by using **-ing, -ed** and **infinitive** clauses.

Here are some more examples:

1. Reason
   The bearing was left to cool down, because it was too hot.
   The bearing was left to cool down, being too hot.

2. Qualification
   The ship managed to reach port, though she was badly damaged.
   The ship managed to reach port, though badly damaged.

3. Time (*after*)
   After the seamen had finished on deck, they went below.
   Having finished on deck, the seamen went below.

4. Time (*when*)
   When the valve has been cleaned, it will work properly.
   (When) cleaned, the valve will work properly.

5. Time (*since*)
   Since I have taken this medicine, I have felt better.
   Since taking this medicine, I have felt better.

6. Advice
   The man you should ask is the Boatswain.
   The man to ask is the Boatswain.

7. Condition
   If the machinery was oiled, it would work better.
   Oiled, the machinery would work better.

*Exercise* 1. *Try reducing these sentences, using the above examples as a guide:*

(a) Because the Lieutenant was the senior officer left alive he took command of the vessel.
(b) Although the radio was damaged, it still functioned.
(c) After the seamen had painted the port side, they started on the starboard side.
(d) When the ship is repaired, she will be put up for sale.
(e) Since he has joined the Royal Navy, he has visited many countries.
(f) The book you should read on that subject is in the library.
(g) If it was painted, it would look as good as new.
(h) As the ropes were wet, they were left on the deck to dry.
(i) They continued rowing, though they were exhausted.
(j) When it has been opened, the container should be emptied immediately.

Sometimes the link between two statements is not made explicit: either a general purpose link such as **and** is used, or nothing at all.

*Exercise* 2. *Rewrite these sentences using a connecting word you have already studied to make the relationship between the two clauses clear. The first has been done for you:*

(a) He loaded the gun and took careful aim.
He loaded the gun, then took careful aim.
(b) He had no money and had to walk into town.
(c) The ship slowed down; the fog had become thicker.
(d) Take this medicine; you will feel better.
(e) They searched carefully; there was no-one there.
(f) He washed the dishes and dried them.
(g) Work hard and you will pass your examinations.
(h) There is a choice of red wine and white wine.
(i) He ran on deck; the alarm bells were ringing.
(j) He tried hard and just failed.

# READING COMPREHENSION

## (B) Communications at sea

*Read through this passage using the techniques summarized in the first part of this unit:*

Communications at sea are essential for the efficient and safe running of a ship. They take place within the ship herself, between the ship and other ships, between the ship and shore stations and sometimes between the ship and aircraft. Communications can be made over different distances and using methods ranging from the simplest to those using the most sophisticated radio technology.

Communications within the ship are done by an internal telephone system. Voice pipes are also used. Engine orders are passed from the bridge to the engine room by means of the ship's telegraph. Messages can also be given to the ship's company through a loudspeaker system. Very large ships had docking telegraphs, usually at the bows and stern. These were used when the vessel was being moored alongside. Nowadays VHF communication is more common.

Communications over relatively short distances can be made by visual or sound signals. Visual signals can be sent by using flags or an Aldis lamp. An Aldis lamp is an electric lamp used for flashing messages in Morse Code. The traditional method of signalling from one ship to another was by using flags. There are different coloured flags for each letter of the alphabet. There are also pennant-shaped flags for numbers, and a long pennant, known as an 'answering' or 'code' pennant. Three other flags, which are burgee-shaped, are known as substitutes. These show that the flag or pennant is being repeated. Besides standing for a letter of the alphabet, each flag, when hoisted alone, has another meaning. For example, the 'W' flag also means: 'I require medical assistance'. Flags can also be hoisted in combinations of two, three or four, all having a particular meaning. Sound signals can be made with the ship's siren, whistle or bell. These are used in fog and similar circumstances when visual signals cannot be seen. The number of blasts signifies that the ship is manoeuvring in a certain way. In emergencies, rockets and flares are used to signal distress and to acknowledge such signals.

**A radio room (*Unit XII*)**

Communications over long distances are sent by radio. Radio communications on board are the responsibility of the Radio Department, although with the introduction of VHF an increasing amount of radio work is being done by the Deck Officer, and in some ships the Radio Officer has been phased out completely. The Radio Officer is responsible to the Master for the efficient operation and maintenance of the communication equipment on board. In most cases he is employed not by the ship's owner, but by a marine radio company. This company pays him, decides which ship he will serve in, arranges his leave, and provides promotional opportunities. The number of radio officers carried on board depends on the statutory radio service laid down for each class of ship. Cargo ships of 1600 gross registered tons and above when fitted, as is normal, with radiotelegraph auto-alarm apparatus (A/A), are required to carry only one radio officer.

On a ship carrying only one radio officer, he keeps a radio watch of eight hours a day, divided into four two-hourly periods. Where there are two radio officers, they keep watch for sixteen hours a day between them, spread over a twenty-four hour period, with four two-hour breaks when neither is on watch. With three or more radio officers, the radio watch is continuous throughout twenty-four hours, each radio officer keeping two four-hour watches with two eight-hour breaks.

A large part of the Radio Officer's watchkeeping duty is taken up with transmitting and receiving radio telegrams on ship's business, navigational and weather messages, time signals for checking the ship's chronometer, social telegrams between the ship's crew or passengers and friends and relatives ashore, and press reports. Many of these messages are handled by radio-telegraphy in Morse Code, but in an increasing number of ships, radio-telephony is being used. This allows the Master and crew members to communicate verbally with the shore.

During watchkeeping periods a continuous watch is kept on the international distress frequency for calls for help or assistance. Listening is intensified during two three-minute periods every hour, when all communications other than distress signals or urgent traffic must stop. When a radio officer goes off watch, the international distress frequency is guarded by an automatic alarm device, which is designed to respond to internationally recognized signals and to actuate alarm bells which alert radio officers to distress calls from other ships.

On a few ships, the Radio Officer serves as a Radio and Electronics Officer and he looks after a wide range of electronic equipment on board, both on the bridge and in the engine room, in addition to the equipment belonging to the radio room.

# APPLIED TERMINOLOGY

## (A) Terms relating to radiotelephone procedures

When transmitting plain language or code, the **Phonetic Alphabet** and **Figure Code** are used.

## The phonetic alphabet

(*Note:* The syllables which are emphasized are underlined.)

*Exercise* 1. *Listen to the recording of the alphabet, look at the table below and repeat the alphabet.*

| Letter | Code | Letter | Code |
|---|---|---|---|
| A | Alfa | N | November |
| B | Bravo | O | Oscar |
| C | Charlie | P | Papa |
| D | Delta | Q | Quebec |
| E | Echo | R | Romeo |
| F | Foxtrot | S | Sierra |
| G | Golf | T | Tango |
| H | Hotel | U | Uniform |
| I | India | V | Victor |
| J | Juliet | W | Whisky |
| K | Kilo | X | X-ray |
| L | Lima | Y | Yankee |
| M | Mike | Z | Zulu |

*Exercise* 2. *When you think you know the Phonetic Alphabet, test yourself by covering the words and saying the letters.*

## The figure code

If there are language difficulties, numerals are pronounced as shown in the table below.

*Exercise* 3. *Listen to the recording of the Figure Code, look at the table below and repeat the figures:*

| Figure | Code word |
|---|---|
| 0 | Nadazero |
| 1 | Unaone |
| 2 | Bissotwo |
| 3 | Terrathree |
| 4 | Kartefour |
| 5 | Pantafive |
| 6 | Soxisix |
| 7 | Setteseven |
| 8 | Oktoeight |
| 9 | Novenine |
| Full stop | Stop |
| Decimal point | Decimal |

*Exercise* 4. ***When you think you know the Figure Code, test yourself by covering the code words and saying them from the figures.***
(*Note:* each syllable should be equally emphasized.)

# (B) Measurement

## *(xii) Frequency*

**Frequency** is the number of times an oscillation repeats itself in one second. It is measured as the number of cycles per second (c/s), though it is usually now expressed in hertz (Hz) (one hertz equals one c/s).

*Study the table below showing measurement of frequency in hertz:*

| Term | Cycles per second | Equivalent |
|---|---|---|
| 1 hertz (Hz) | 1 | |
| 1 kilohertz (kHz) | 1,000 | 1,000 Hz |
| 1 megahertz (MHz) | 1,000,000 | 1,000 kHz |
| 1 gigahertz (GHz) | 1,000,000,000 | 1,000 MHz |

Frequencies range from very low frequencies (vlf) to extremely high frequencies (ehf). Other frequencies include: high frequencies, super-high frequencies, low frequencies, very high frequencies, medium frequencies, ultra-high frequencies.

*Exercise* 1. *Arrange the above frequencies in order and write them in the left hand column of the table below. Add the abbreviations.*

| Frequency band | Frequency range | Typical uses |
|---|---|---|
| very low frequencies (vlf) | 3–30 kHz | .................................. |
| ..............................(...) | 30–300 kHz | .................................. |
| ..............................(...) | 300–3000 kHz | .................................. |
| ..............................(...) | 3–30 MHz | .................................. |
| ..............................(...) | 30–300 MHz | .................................. |
| ultra high frequencies (uhf) | 300–3000 MHz | .................................. |
| ..............................(...) | 3–30 GHz | .................................. |

Radio frequency bands have been allocated by international agreement to various users. These are summarized on the table below.

| Utilization of radio-frequency spectrum | |
|---|---|
| | time signals, long distance |
| 30 kHz | |
| | maritime mobile, navigational aids, fixed |
| 300 kHz | |
| | maritime mobile, broadcasting |
| 3 MHz | |
| | fixed, maritime and aeronautical mobile, broadcasting, amateur |
| 30 MHz | |
| | television and radio broadcasting, radar, maritime and aeronautical mobile, short wave, broadcasting, fixed, mobile |
| 300 MHz | |
| | fixed, mobile, maritime and aeronautical mobile, broadcasting, navigation, meteorological, space, short distance |
| 3 GHz | |
| | Radar, radio relay, navigation, space, satellite communication, experimental |
| 30 GHz | |

*Exercise* 2. *Transfer the information to the table above putting it in the column marked 'Typical uses'. You will need to use abbreviations.*

# GUIDED WRITING

## (A) Description of radio communication system

*Write a description of the basic elements in a radio communication system.*

*Stage* 1. *Write an introductory paragraph saying what the system consists of. Use the diagram and notes below.*

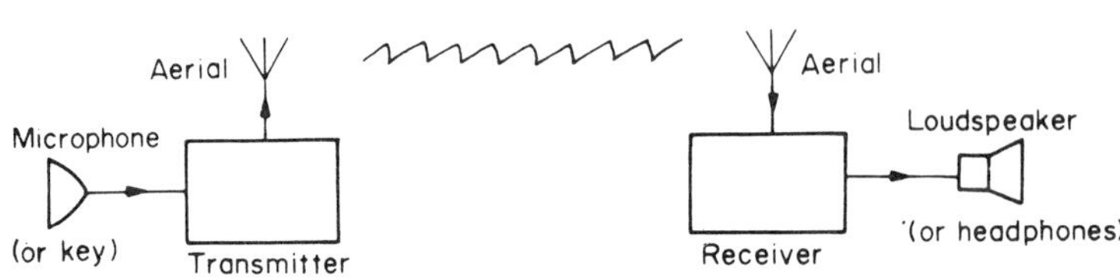


The microphone produces an electrical signal.
This can also be done by a Morse key.

The radio transmitter generates radio-frequency waves.
These waves vary in accordance with the electrical signal.
The transmitting aerial radiates the waves into space.
The receiving aerial receives the radiated waves.
The radio receiver amplifies the signal.
The radio receiver reproduces the original electrical signal.
The loudspeaker converts the electrical signal into sound.
This can also be done by headphones.
The original sound is reproduced.

*Stage* 2. *Read through the following statements and divide them up into groups according to subject matter. Join each group of statements together into sentences to form paragraphs. Make any changes you think necessary:*

The information to be transmitted is superimposed on radio waves.
This is done by the transmitter.
It uses the signal from the Morse key or microphone to do this.
The transmitter generates the radio waves at a definite frequency.
The waves are conveyed to the transmitting aerial for radiation.
The information to be transmitted must first be converted into an electrical signal
In radio-telegraphy this is done using a Morse key.
In radio-telephony this is done using a microphone.
The receiving aerial picks up the radio signal.
It picks up a mass of other signals as well.
The receiver must do a number of things.
It must select the desired signal.
It must amplify the desired signal.
It must reproduce the original electrical signal.
This is then sent to headphones or a loudspeaker.

*Stage* 3. *Use the diagram to help you sort out your paragraphs into a logical order. Put them together with your introductory paragraph to form your description. Write transitional sentences and a summary sentence, if you think they will improve your description.*

# NOTE-TAKING PRACTICE

## Phonetic alphabet

When there are language difficulties the nature of distress can be sent in code.

*Practice* 1. *Listen to the following code signals taken from the International Code of Signals and read out using the Phonetic Alphabet. Write them in the column maked 'Code letters'.*

| Code letters | Meaning of signal |
|---|---|
| ........................ | I must abandon my vessel. |
| ........................ | Aircraft is ditched in position indicated and requires immediate assistance. |
| ........................ | I require immediate assistance. |
| ........................ | I require immediate assistance. I am on fire. |
| ........................ | I am sinking. |
| ........................ | I have collided with surface craft. |
| ........................ | I am proceeding to your assistance. |
| ........................ | Your distress signals are understood. |
| ........................ | Repeat the distress position. |

A ship's position can also be sent using the Phonetic Alphabet and the Figure Code. This can be done as follows:

1. By bearing and distance from a landmark
   (a) Code letter A (Alfa) followed by a three-figure group for the ship's **true** bearing from landmark
   (b) Name of landmark
   (c) Code letter R (Romeo) followed by one or more figures for distance in nautical miles.
   e.g. A 015 Ushant R 40.

2. By latitude and longitude

Latitude: (a) Code letter L (Lima) followed by a four-figure group (two figures for degrees, two figures for minutes)

(b) Either N (November) for latitude North, or S (Sierra) for latitude South.

Longitude: (a) Code letter G (Golf) followed by a five-figure group (three figures for degrees, two figures for minutes)

(b) Either E (Echo) for longitude East or W (Whisky) for longitude West.

e.g. L5425 N G 01633 W.

*Practice* 2. *Listen to these distress messages and write down the details by completing the outlines below:*

(a) Ship's name: ..........................................................................

Position: ..........................................................................

Nature of distress: ..........................................................................

(b) Ship's name: ..........................................................................

Position: ..........................................................................

Nature of distress: ..........................................................................

## **Lecture:** Authorities giving aid to ships and aircraft

Now listen to the following lecture and take as full notes as possible, using the techniques practised throughout this course. Don't forget to give your notes a title and to label each paragraph clearly.

# GUIDED WRITING

## (B) Authorities giving aid to ships and aircraft

*Use the notes that you have taken during the lecture in the previous section to write an essay on the different authorities which will help ships and aircraft in distress around the coasts of the United Kingdom.*

# UNIT XIII

# RADIO COMMUNICATIONS

## READING COMPREHENSION

### (A) Radio communications

At nautical college you will be given a number of handouts, on which lessons and tests will be based.

*Below is the type of handout you can expect to get from the Radio Department. Notice the short topic paragraphs, often written in table form; the compressed style of writing with a lot of information in it; the use of abbreviations, brackets and symbols.*

A communication system enables information to be passed from a source, through a medium, to a destination. Between a ship and a land station that medium will be radio waves. From the land station the medium is usually land line, i.e. the internal telephone and telecommunications system. Messages can be sent in either Morse code or speech. If Morse is used, then wireless-telegraphy (W/T) is used. When speech is used, then radio-telegraphy (R/T) or VHF are used. With R/T the receiving operator has to convert the Morse code into plain language. When transmitting a message by R/T or VHF, the International Phonetic Alphabet must be used to avoid confusion when there are language difficulties.

Transmissions from all stations must contain an identity signal by which the station making the transmission can be recognized. This signal must be used at all times when communicating between stations. Such identification must take the form of International Callsigns or other internationally agreed forms of identity. In radiotelephony, coast (land) stations normally identify themselves by using their geographical name, followed by the word **Radio**, e.g. **Nitonradio**, **Angleseyradio**. Ship stations normally identify themselves by the name of the ship, but the International Callsign assigned to the ship may be used, for example, when two or more ships bear the same name, or where pronunciation/ language difficulty could occur.

The three marine frequencies used for Distress, Call and Reply purposes are:

(a) 500 khz for telegraphy (Morse)
(b) 2182 khz for telephony (speech)
(c) 156.8 Mhz for telephony (speech)

During hours of service all ships and coast stations maintain a listening watch on at least one, if not all, of the above frequencies for information regarding messages for individuals, general messages, and distress, urgency and safety messages. The obligation to accept distress calls and messages is absolute for all stations and such messages must be given priority and dealt with immediately.

In order to increase the safety of life at sea, and over the sea, all stations of the maritime mobile service normally keeping watch in the authorized band between 405 – 535 khz must, during their hours of service, ensure watch on the International Radiotelegraph Distress/Calling frequency 500 khz is kept for three minutes, **twice each hour**, at h15 – h18 and at h45 – h48. Those normally keeping watch in the frequency band 1605 – 2850 khz must, during their hours of service, ensure watch on the International Radiotelephone Distress/Calling frequency 2182 khz is kept for three minutes, **twice each hour**, at h00 – h03 and at h30 – h33. These times are referred to as **silence periods** and must always be observed. During these three-minute silence periods all transmissions on these frequencies must stop in order that possible weak signals of a distress call can be heard. Transmissions between the frequencies 485 – 515 khz, and between 2173.5 – 2190.5 khz, must cease during silence periods, except for distress transmissions.

Where VHF radiotelephony equipment is installed, a continuous listening watch is maintained on the International Distress/Calling frequency 156.8 Mhz, referred to as Channel 16. No silence periods are allocated to this frequency.

The Classification of Emissions refers to the different types of transmission. These are symbolized as follows (only the more common types are described):

A1A TELEGRAPHY by on/off keying, unmodulated full carrier.
A2A TELEGRAPHY by on/off keying, amplitude modulated double sideband full carrier.
H2A TELEGRAPHY by on/off keying, amplitude modulated single sideband full carrier.
A3E TELEPHONY amplitude modulated double sideband full carrier.
H3E TELEPHONY amplitude modulated single sideband full carrier.
R3E TELEPHONY amplitude modulated single sideband reduced carrier.
J3E TELEPHONY amplitude modulated single sideband suppressed carrier.
F3E TELEPHONY frequency modulated carrier. VHF 156 – 174 Mhz Marine.

Note. 1 A3E and H3E are not permitted, except on 2182 khz for, and by, distress equipment for safety purposes.

2 H3E may be used on 2182 khz for calling purposes, but **not** on any other frequency used for calling purposes.

3 When establishing communication on working frequencies, either R3E or J3E must be used.

## *Short, medium and long range communications*

In general 'short' range communication refers to VHF (156 – 174 Mhz) and the radius of such contact is 30 – 45 miles. The main use is Telephony (F3E). 'Medium' range communication can refer to MF Telegraphy (W/T) in the 405 – 535 khz, or MF Telephony (R/T) in the frequency band 1605 – 4000 khz. The radius of contact is roughly between 150 – 300 miles, depending on local propagation conditions, transmitter power and the aerial system used. 'Long' range communication refers to the HF band of frequencies, 4000 – 25000 khz for W/T (A1A only), and 4000 – 23110 khz for R/T (R3E or J3E only). 'Long' range is used for distances exceeding 300/400 miles.

Certain VHF channels are designated for particular purposes. These are given in the International Telecommunication Union (ITU) 'List of Coast Stations'.

| | | |
|---|---|---|
| Distress, Safety and Calling | | Channel 16 |
| Intership | first choice | Channel 6 (Others listed in VHF allocation guide) |
| Port operations | | Channel 10 (also 12, 14) |
| Public correspondence | main choices | Channels 24, 25, 26, 27, 28 |
| On-board | | Channels 15, 17 (on low power) |
| Coast Radio Stations | | channels listed in ITU List of Coast Radio Stations or as specified by the CRS. |

Note. 1 Port Operations – ship movement and Pilot Services
Port Operations channels (found in 'List of Coast Stations') are restricted to various port operation services, the movement and safety of ships, and, in emergency, to the safety of persons. **No private or public correspondence** messages can be passed over these channels.

2 Public Correspondence – any telecommunication which the offices and stations must, by reason of their being at the disposal of the public, accept for transmission.

3 On-board Communications
These are restricted to communications on-board the ship itself, mooring/unmooring and cargo operations; or between the ship and its liferafts; or between a group of vessels during towing operations.

*Questions related to handouts tend to ask you to look for information in the text.*
*e.g. State the three marine frequencies used for Distress, Call and Reply purposes. Give the silence periods which must be observed by those keeping watch in the frequency band 1605 – 2950 khz.*

# GRAMMAR

## (A) Message types

In the **Seaspeak** Reference Manual, which gives essential English for VHF use, messages are divided into only seven types based on their function. These seven types are listed below with examples:

1. **Question** e.g. (a) What is your position?
   (b) Is visibility expected to change?

2. **Instruction** e.g. (a) You must stay clear of the fairway.
   (b) Do not overtake.

3. **Advice** e.g. (a) Advise you pass astern of me.

4. **Request** e.g. (a) I require a tug.
   (b) Please send medical assistance.

5. **Information** e.g. (a) The pilot is waiting now at position ....
   (b) My ETA at East Pier is one-six-three-zero local.

6. **Warning** e.g. (a) I am manoeuvring with difficulty.

7. **Intention** e.g. (a) I intend to reduce speed, new speed: five knots.

Note: Here **Instruction** refers to commands, and **Advice** to suggestions. A **Warning** contains information of critical importance to the safety of the vessel, and an **Intention** indicates immediate operational intentions.

*Exercise* 1. *In order to understand a message completely it is necessary to know its function. See if you can identify the function of these messages by writing:* **question, instruction, advice, request, information, warning** *or* **intention** *in the space provided, as appropriate.*

(a) ..............: Buoy number two-six is unlit.
(b) ..............: What are my berthing instructions?
(c) ..............: No vessels are at the anchorage.
(d) ..............: Go to berth number eight.
(e) ..............: What is your draught aft?
(f) ..............: Advise you keep your present course.
(g) ..............: I expect to be underway within period: two hours.
(h) ..............: Please provide fire-fighting assistance.
(i) ..............: Push on my port bow.
(j) ..............: The vessel ahead of you is stopping.

In the Seaspeak system a message should always be initiated by a 'message marker', i.e. one of the seven words above denoting message type. This will increase the probability of the message being understood. The system also uses 'reply markers', which correspond as follows: **question/answer, instruction/instruction-received, advice/advice-received**, etc.

e.g. **question:** What is your ETA at the harbour entrance?
**answer:** My ETA at the harbour entrance is: time: one-four-zero-zero GMT.

With the other markers, the message is quoted back in full after the 'reply marker' with any necessary changes made.

e.g. **intention:** I intend to reduce speed, new speed: five knots.
**intention-received:** You intend to reduce speed, new speed: five knots.

## (B) Message patterns

Seaspeak recommends that a message and a reply should take the following patterns:

1. **Questions** These should be restricted to three types:
   (i) Questions beginning with 'Wh' question words or 'How' e.g. What ..., When ..., Where ..., How many ..., etc.

   (ii) Questions which give alternatives, e.g. Are you loading or unloading?

   (iii) Questions requiring a 'yes' or 'no' answer.

*Note.* Questions of the '......., isn't it?', '........, won't you?' type, and those formed by tone of voice are strongly advised against.

2. **Answers**

When a yes or no answer is required these take the form of: **Positive, Negative** or **I don't know**, as follows:

e.g. **question:** Are there any survivors?
**answer:** Positive, there are four survivors.
or
**answer:** Negative, there are no survivors.
or
**answer:** I don't know if there are any survivors.

3. **Instructions**

These should use the Imperative form,
e.g. Stop ... Turn ... Do not anchor ....

Agreement or disagreement with the instruction is indicated by saying 'Positive' or 'Negative' after quoting the command. After a negative a reason may be given.

e.g. **Instruction:** Stop your engines.
**Instruction-received:** Stop my engines, negative, reason: the tide is too strong.

4. **Advice**

This signals suggestions and also uses the Imperative or 'Advise you ...'.

5. **Requests**

Requests for items to be delivered begin: 'Please deliver ...'; 'Please supply ...'; 'Please send ...'; 'Please provide ...'; or 'Please confirm ...' (if seeking confirmation of information).

6. **Information & warnings**

These take the form of statements and often use the Simple Present and Continuous Present tenses. See (A) Message Types 6 & 7.

7. **Intentions**

Although in everyday English intentions can be expressed using 'going to', Seaspeak recommends the pattern: 'I intend to ...'.

8. **Reasons**

If a reason is given with any of the above, Seaspeak recommends that the single word 'Reason' should be used rather than words such as, 'because', 'as', 'so that', 'in order to', etc.

e.g. **advice:** Please keep clear of me, reason: my steering-gear is defective.

*Exercise* 1. *Use the 'message markers' and the message patterns described above to complete these conversation extracts, which are based on those found in the Seaspeak Reference Manual.*

(a) ........................ Dredging operations are completed in the fairway. Over.

**Information –** ....................: ...............................

....................: ............................................... Over.

(b) ....................: ............................................... over.

.................... – **received:** You intend to anchor now. Over.

(c) **Question:** Is buoy number: two-three in the correct position? Over.

....................: Negative, ........................................

............................................... Over.

(d) **Request:** ........ welding equipment, ..... urgent repair.

....................: We have qualified welders on board. Over.

(e) ....................: Please proceed to Avonport, ..................

............................................... Over.

**Advice –** ........... ..............................................

.......... , reason: there is cargo for me. Over.

(f) ....................: A strong easterly wind is blowing at the berth. Over.

....................– .................... : ...............................

............................................... Over.

(g) ....................: Turn starboard immediately. Over.

....................– ....................: ...............................

.......................................... , negative,

........................ ......: sailing vessel on my starboard bow. Over.

*Exercise 2. Rewrite these sentences in the Seaspeak format.*

(a) Large vessel leaving. Keep clear of the approach channel.

..........................................................................................

(b) I will attempt rescue by Breeches-buoy.

..........................................................................................

(c) At what time do you expect to arrive at the harbour entrance?

..........................................................................................

(d) You did say your length was two hundred metres, didn't you?

..........................................................................................

(e) It would be better for you to pass ahead of me as I am slowing down.

..........................................................................................

(f) I'm afraid that shore based radar assistance is not available.

..........................................................................................

(g) Okay. Thanks for the warning. I'll keep a look out for wreckage.

..........................................................................................

# READING COMPREHENSION

## (B) VHF procedures

*Read through the text below, which is designed to give you further practice in reading handouts and manuals. With the description of the exchange procedures layout is vitally important for clarity.*

When using VHF (156 – 174 Mhz), call/reply is normally made on Channel 16, but other arrangements may be available on other channels.

Prior to making a call it is first necessary to ensure there is no distress, urgency or safety traffic in progress and that another communication is not being interfered with. A call must **not** exceed one minute, but may be repeated at three-minute intervals.

Here are two possible exchanges. The first is an exchange procedure between two ships taken from Seaspeak. The second is an exchange procedure between a ship and a coast station using standard procedure.

An outline of an intership exchange using Seaspeak:

| Station making the call | Station responding to the call |
| --- | --- |
| 1. *Initial call*<br>Address.<br>(Name/Callsign of station called, said twice)<br>Identification.<br>'This is ...'<br>(Name/Callsign of calling station, said twice)<br>Indication of calling VHF channel.<br>'On VHF channel ...'<br>'Over.' | |
| | 2. *Response to call*<br>Address and identification. 'Over.' |
| 3. *Indication of working VHF channel*<br>Address and identification.<br>Indicate working VHF channel.<br>'Switch to VHF channel ...'<br>'Over.' | |
| | 4. *Agreement of working VHF channel*<br>Address and identification.<br>Agree working VHF channel.<br>'Agree VHF channel ...' |

(Both ships now change to the agreed working channel.)

| Station making the call | Station responding to the call |
| --- | --- |
| 5. *Message*<br>Address and identification.<br>Message.<br>'Over.' | |
| | 6. *Response to message*<br>Address and identification.<br>Respond to message.<br>'Over.' |
| 7. *End of transmission*<br>Address and identification.<br>Read back response to message.<br>'Out.' | |

An outline of a ship-to-shore exchange using standard procedure:

| Station making the call | Station responding to the call |
| --- | --- |
| 1. *Initial call*<br>Address. | |

(Name/identity of station called, said twice)
Identification.
(Name/identity of calling station, said twice)
'This is ...'
Reason for call, giving working channel.
'Over.'

2. *Response to call*
Address and identification.
'Romeo Channel ... up'

(Both ships now change to the agreed working channel.)

3. *Indication of readability*
Address and identification.
'How do you read me?'
'Over.'

4. *Indication of readability*
Address and identification.
'Read you strength ... go ahead.'
'Over.'

5. *Message*
Address and identification.
'Message begins ...'
'Message ends.'
'Over.'

6. *Response to message*
Address and identification.
'Received message number ...'
'Nothing for you.'
'Over.'

7. *End of transmission*
Address and identification.
'Romeo nothing for you.'
'Out.'

If a station receives a call, but is uncertain that the call is actually intended for it, it must wait until the call has been repeated and understood.

In ship-to-shore communications, except in the cases of distress, urgency or safety, the communications are controlled by the coast station. In ship-to-ship communications the ship which is calling controls the exchange. If a coast station finds it necessary to interrupt, both ships must comply with any instructions given by the shore station.

*Now complete this conversation exchange using the information above and the language work in the Grammar section to help you. The exchange uses the Seaspeak procedure.*

Station making the call — Station responding to the call

1. 'Polar Star, Nine Victor Alpha Tango;

.......................................... ;

This is Rambler, Golf Xray Xray Xray

..........................................

................................ one-six.

......'

2. ' ..........................................

............. Polar Star, Nine Victor

Alpha Tango.

......'

3. ' .................... This is Rambler.

................................ zero-six.

Over.'

4. ' ..........................................

Agree ..................................

......'

5. ' ..........................................

......... : What is your destination?

Stay on.

......'

6. ' ..........................................

.........: ....................... Dakha.

......'

7. ' ..........................................

Understood, Dakha.

................ : Dakha pilot services

suspended, reason: gales.

......'

8. 'Rambler. This is Polar Star.

..........–..........: ....................

..........................................

Over.'

9. ' ..........................................

Nothing more.

......'

# APPLIED TERMINOLOGY

## (A) Terms relating to VHF communications

### *Standard phrases*

*Study the table below, which shows some of the standard phrases used in VHF communications. These are taken from Standard Marine Navigational Vocabulary (IMCO) and Seaspeak (Pergamon).*

| | |
|---|---|
| *Making and maintaining contact*<br>Calling ...<br>This is ...<br>How do you read?<br>I read ... (1–5)*<br>Stand by on VHF channel ...<br>Standing by on VHF channel ...<br>Over<br>Out<br><br>*Conversation controls*<br>Please read back<br>I read back<br><br>*Announcements*<br>Message for you.<br>Pass your message. | *Clarification*<br>Mistake.<br>Correction.<br>Please spell.<br>I spell.<br>Say again all before/after ...<br>I say again all before/after ...<br>Please speak in full.<br>Please speak slowly.<br>Please use Standard Marine Navigational vocabulary.<br>Please use International Code of Signals.<br>Please use Seaspeak.<br><br>*Channel switching*<br>Switch to/Change to VHF channel ...<br>Stand by on channel ... |

**Note*. In Seaspeak: 1 = unusable, 2 = poor, 3 = fair, 4 = good, 5 = excellent.

## *VHF distress, urgency and safety messages*

*Study this diagram which tells you when to use distress, urgency and safety procedure. The procedures here are based on those recommended in Seaspeak.*

| Type of procedure | Call | Situation |
|---|---|---|
| Distress | Mayday | To be used when a ship or aircraft is threatened by grave and imminent danger likely to involve loss of life, and requests immediate assistance. |
| Urgency | Pan-Pan | To be used when the message contains urgent information concerning the safety of a ship, aircraft or other vehicle, or the safety of a person. |
| Safety | Sécurité (pronounced Say-cure-e-tay) | To be used when the message contains an important navigational or meteorological warning. |

*Exercise* 1. *Look at these messages and decide whether distress, urgency or safety procedures should be used.*

(a) 100 miles southwest of Landsend engine broken down. Require tug assistance. .........

(b) From Harbourmaster Falmouth yacht **Belinda** overdue on voyage from **Cromer** to **Falmouth** left Cromer 10th March report any sightings to **Lloyds London**. .........

(c) Diving operations in progress at ninian north platform ship movement within a radius of 500 metres prohibited until further notice. .........

(d) Ship holed in engine room. Sinking. Request immediate assistance. .........

(e) Gale warning southwest gale force winds imminent in sea areas Dover Thames. .........

(f) 52.30 North 16.20West fire in number two hold ships in vicinity please stand by. .........

The procedure for a distress or urgency transmission is as follows.

Distress call (3 times)
'This is ...'
Name of ship (3 times)
Distress call (once)
Name of ship and Callsign (once)
Position
Nature of distress

Kind of assistance required
'Over.'

*Exercise* 2. *Now complete these two transmissions, taken from the Seaspeak Manual, by selecting the appropriate phrases from the jumbled list below.*

| | |
|---|---|
| (a) ' .................................. | (b) 'Pan-Pan Pan-Pan Pan-Pan |
| ........... | ........... |
| Rattler Rattler Rattler | ....................................... |
| ........... | ............ |
| ................................ | Vega, Seven Victor Alfa Tango |
| Position: latitude: five-zero | ....................................... |
| degrees three-zero minutes | ....................................... |
| North, longitude: zero-three- | ....................................... |
| nine degrees two-zero minutes | ....................................... |
| West. | ....................................... |
| ....................................... | ....................................... |
| ....................................... | ....................................... |
| ...........' | Over.' |

This is / Require tow / Pan-Pan / Rattler, Golf Xray Xray Xray / Collision with iceberg, sinking / Mayday Mayday Mayday / Position: bearing zero-nine-zero degrees true, distance one-five miles from Ras Sarkan / Mayday / Lost propeller / Over / Vega Vega Vega / This is / Request immediate assistance.

The procedure for a safety transmission is as follows.

1. Safety call (3 times)
2. 'All ships' (may be repeated)

1–6. Repeated.
7. Safety message.

3. 'This is ...'
   Name of station (3 times)
4. Safety call (once)
5. Name of station
6. Phrase indicating content of message to follow.
7. 'Switch to VHF channel ...'
8. 'Over'
8. 'I say again.'
   Repeat safety message.
9. 'Out.'

*Exercise* 3. *You are* **Nitonradio** *and you wish to make a safety message broadcast to all ships that there is a military exercise involving darkened ships, submarines and aircraft taking place in sea area* **Delta** from 0800 hours GMT, March 3rd to 2200 hours GMT, March 7th.

Using the above procedure outline write out the complete transmission.

## (B) Measurement

### *(xiii) Transmission of measurements and quantities*

1. Numbers.

Except when there are language difficulties and the Figure Code is used (see page 176), numbers are pronounced as in everyday English, but with modifications for a few numbers.

*Exercise* 1. *Listen to your teacher reading out the numbers and repeat them, first with and then without the book.*

| Figure | Spelling | Pronunciation | Figure | Spelling | Pronunciation |
|---|---|---|---|---|---|
| 0 | zero | zero | 6 | six | six |
| 1 | one | wun | 7 | seven | seven |
| 2 | two | too | 8 | eight | ait |
| 3 | three | tree | 9 | nine | niner |
| 4 | four | fower | 1000 | thousand | tousand |
| 5 | five | fife | | | |

Numbers should be read digit by digit.

e.g. 25 two-five
609 six-zero-nine

| | | |
|---|---|---|
| | 45,136 | four-five-one-three-six |
| **but** | 45,000 | four-five-thousand |

2. Measurement.

Using Seaspeak, measurements are given in the following order:

(i) what is being measured (length, draught, speed, etc.);
(ii) the numbers (one-four-six, etc.);
(iii) the units of measurement (metres, knots, metric tonnes, etc.)

e.g. length: one-four-six metres
speed: one-zero knots
My draught forward is: eight metres

3. Quantity.

A quantity of items (e.g. referring to stores, cargo, etc.) is expressed by the word **quantity** followed by the number and then the name of the item.

e.g. quantity: five acetylene cylinders
quantity: four-zero metric tonnes fresh water
quantity: two-zero-zero litres lube oil

When making a request these are used with verbs such as: **to supply, to deliver, to provide, to send**, etc.

Using Seaspeak you would say: '**Request:** Please deliver: quantity: five acetylene cylinders.'

*Exercise 2. Using SEASPEAK how would you make the following requests:*

(a) You want 500 tonnes of fresh water delivered to berth no. 18.
(b) You want to be supplied with welding equipment and two sets of gas bottles.
(c) You want 10 litres of clear varnish and 50 litres of deck paint, green.
(d) You need a replacement scanner motor for a Decca Radar Transar 809. The part number is 110VDC2504Z.

# GUIDED WRITING

## (A) Keeping the radiotelephone log

In accordance with shipping regulations a radiotelephone log must be kept on board ship. This log is normally kept by the Radio Officer. The log contains a complete record of each radio watch, recording full details of *all* calls and signals: the exact time sent or received, the stations involved, the frequencies used and their content; in addition to other information. The log must be kept at the place where the radio watch is maintained and it must be available for inspection.

In order that a ship may receive messages as quickly as possible, the 'local' coast station must be given details of the ship's movements. This is known as a TR (Transit Record), said 'Tango Romeo', and comprises:

1. Ship's name and call sign.
2. Ship's position, which may be given by:
   (i) latitude and longitude, or
   (ii) distance and bearing from coast station, or
   (iii) well-known geographical point/ location.
3. Ship's next port of call.

The TR must be transmitted in the above order and sent on first contact with a coast station, prior to commencing traffic; when entering the service area of a coast station; prior to entering port; and immediately on leaving port. The particulars given by the ship are 'filed' at a central office so any incoming messages/ telephone calls, etc., can be routed to the coast station in whose area the ship is.

*Study the extract below of part of a ship's radio log. What information can you get from it?*

S.S. ..... ESSEX COAST GMML
M.V.

DIARY OF THE RADIOTELEGRAPH SERVICE

| DATE AND TIME G.M.T. | STATION FROM | STATION TO | FULL DETAILS OF CALLS, SIGNALS AND DISTRESS WORKING, ETC., AS PRESCRIBED BY RULE 19 OF THE MERCHANT SHIPPING (RADIO) RULES, 1965 | FREQUENCY |
|---|---|---|---|---|
| 12.3.84 | | | QTO Aberdeen bound Leith | |
| 0830 | | | On watch J. Smith. Battery tested – fully | 2182 |
| | | | charged condition | |
| 30/33 | | | Silence period observed | " |
| 35 | | | L.S.W. ON | " |
| 35 | GMML | GND | QTC 1 TR. QSW 2016 K | 2381 |
| 36 | GND | GMML | R 2016 1sn 1856 K // R up | 1792 |
| 37 | GMML | GND | HW? K | 2016 |
| 37 | GND | GMML | QRK 5 QRV K | 1856 |
| 39 | GMML | GND | TR ESSEX COAST/GMML QTO ABERDEEN BND LEITH K | 2016 |
| 39 | GND | GMML | R TR K | 1856 |

What was the name of the vessel and its call sign?

..........................................................................................

Where was the vessel bound for?

..........................................................................................

What was the name of the Radio Officer on watch?

..........................................................................................

What time did he begin his watch?

..........................................................................................

Which silence period did he record observing?

..........................................................................................

When writing up the radio log, abbreviations are used where possible. The 'Q' Codes are used and also abbreviations, some of which are shown below.

Abbreviations

| | | | |
|---|---|---|---|
| ETA | Estimated Time of Arrival | MIN | minutes |
| ETD | Estimated Time of Departure | NIL | I have nothing for you |
| HW? | How are you receiving me? | NX | Navigational Warning |
| HW | I read you ... (1–5) | N | North |
| K | Over | S | South |
| KTS | knots | E | East |
| R | Received | W | West |
| R up | I agree working frequency | WX | weather report |

## *'Q' Codes*

*Now study this table, which gives examples of the 'Q' Code, an internationally recognized code of abbreviations used to overcome language difficulties and reduce the amount of time needed to pass information.*

| | | | |
|---|---|---|---|
| QRB? | How far approx. are you from my station? | QSS? | What working frequency will you use? |
| QRB | The approx. distance between our stations is ... nautical miles. | QSS | I will use working frequency ... |
| | | QSW? | Will you send on this frequency/ on ... khz(Mhz)? |

| | | | |
|---|---|---|---|
| QRD? | Where are you bound for and where are you from? | QSW | I will send on this frequency/ on ... khz (Mhz). |
| QRD | I am bound for ... from .... | QTC? | How many telegrams have you to send? |
| QRJ? | How many radiotelephone calls have you to book? | QTC | I have ... telegrams for you. |
| QRJ | I have ... radiotelephone calls to book. | QTO? | Have you left port/dock? |
| | | QTO | I have left port/dock. |
| QRK? | What is the intelligibility of my signals? | QTP? | Are you going to enter port/ dock? |
| QRK | The intelligibility of your signals is ... (1 – 5) | QTP | I am going to enter port/dock. |
| | | QUA? | Have you news of ... ? |
| QRT? | Shall I stop sending? | QUA | Here is news of ... |
| QRT | Stop sending | N.B. | In R/Tthese are said phonetically. For "?" say "Romeo Quebec". |
| QRU? | Have you anything for me? | | |
| QRU | I have nothing for you. | | |
| QRV? | Are you ready? | e.g. | QTC "QuebecTango Charlie" |
| QRV | I am ready. | | QTC? "QuebecTango Charlie Romeo Quebec" |

*Now use the abbreviations above and the table of 'Q' Codes to help write out these notes in your own words.*

(a) QTC 1 TR QSW 2016 K

.....................................................................................................

(b) R 2016 lsn 1856 K // R up

.....................................................................................................

(c) HW? K

.....................................................................................................

(d) QRK 5 QRV K

.....................................................................................................

(e) TR ESSEX COAST/GMML QTO ABERDEEN BND LEITH K

.....................................................................................................

(f) R TR K

.....................................................................................................

# NOTE-TAKING PRACTICE

## Radio telegrams

*Study the sample of a radio telegram form and the notes below.*

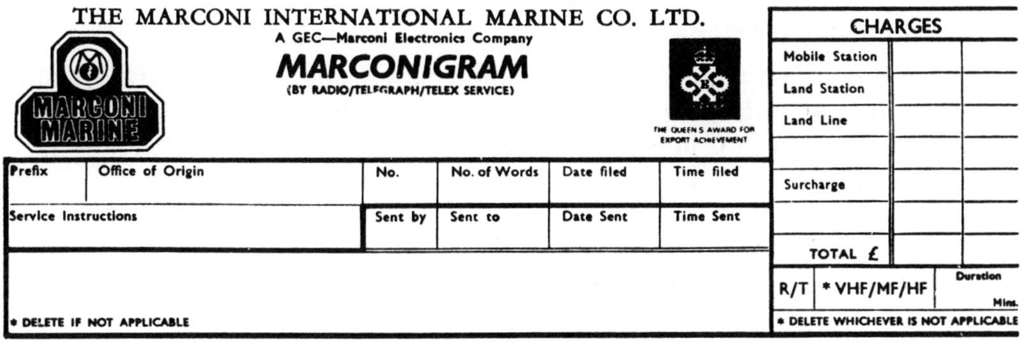
THE MARCONI INTERNATIONAL MARINE CO. LTD.
A GEC—Marconi Electronics Company
MARCONIGRAM
(BY RADIO/TELEGRAPH/TELEX SERVICE)
MARCONI MARINE
THE QUEEN'S AWARD FOR EXPORT ACHIEVEMENT

| Prefix | Office of Origin | No. | No. of Words | Date filed | Time filed |
|---|---|---|---|---|---|
| Service Instructions | | Sent by | Sent to | Date Sent | Time Sent |

* DELETE IF NOT APPLICABLE

| CHARGES | | |
|---|---|---|
| Mobile Station | | |
| Land Station | | |
| Land Line | | |
| | | |
| Surcharge | | |
| | | |
| TOTAL £ | | |

R/T | * VHF/MF/HF | Duration Mins.

* DELETE WHICHEVER IS NOT APPLICABLE

The component parts of a radio telegram are:

1. Preamble (outlined in heavy ink).
2. Service indicator (marked off by = signs).
3. Address.
4. Text.
5. Signature (not compulsory).

1. The Preamble comprises:

(i) Prefix, which indicates a particular type of telegram.
Common prefixes in use are:

| | | | |
|---|---|---|---|
| A | free service advice message | ETAT | Government or State telegram |
| SVC | free service advice message | SLT | Ship Letter Telegram |
| ST | paid service advice message | RTL | Radio Telex Letter |
| RST | reply to paid service advice | MSG | Ship's business telegram |
| OBS | Official Meteorological service advice | P | Private telegram |

(ii) Office of Origin. Where 'handed in'.

(iii) No. of telegram. Numbered separately to each station daily from 0001 hrs.

(iv) No. of words. Includes all that the sender writes, plus any special service required shown by the Service Indicator before the address. No charge is made for the preamble or for = and + signs. There is a fixed minimum charge of 7 words. A word is defined as a group of characters (letters, figures, punctuation signs) separated by spaces. Where a word/expression exceeds 10 characters, it is charged at the rate of 1 word for each 10 characters, or part thereof.

(v) Date filed. Date handed in, indicated by two digits (01-31).

(vi) Time filed. Time handed in, always GMT and indicated by four digits (0001 – 2359).

(vii) Service instructions. Additional information added to a telegram by the office of origin e.g. 'via', 'relayed by', 'redirected from'. Carries no charge. The A/C indicator is inserted here, e.g. GBØ8, which indicates that the Marconi Company settles the account.

2. Service indicator. This denotes a particular type of telegram or Special Service requested by the sender or addressee, e.g.:

| | |
|---|---|
| = RP = | Prepay a reply to any value. |
| = TC = | Have it repeated from office to office during transmission. |
| = ST = | Cancel, alter or amplify a telegram already transmitted. |
| = GP = | Have it delivered to a Poste Restante. |
| = TR = | Have it delivered to a Telegraph Restante. |
| = MP = | Have it delivered to the addressee in person. |
| = URGENT = | Obtain priority over the Telegraph system. |
| = POSTE = | Delivered by ordinary mail from the Telegraph Office. |

3. Address. Requirements for a telegram to an addressee on board ship:
   (i) Name or designation of addressee.
   (ii) Name of ship (and call sign).
   (iii) Name of coast station through which telegram is to be routed.

   Categories of address admitted: full postal; P.O. Box; registered/telegraphic; telephonic; telex; Poste Restante/Telegraph Restante.

4. Text. In plain or secret language in letters of the Roman (English) alphabet.

5. Signature. Not compulsory and may be written in any form.

*Now listen to this conversation exchange between a ship and a coast radio station and take down the message on the radio telegram form below, completing all the details in the appropriate places from the information given.*

THE MARCONI INTERNATIONAL MARINE CO. LTD.
A GEC—Marconi Electronics Company

**MARCONIGRAM**

(BY RADIO/TELEGRAPH/TELEX SERVICE)

MARCONI MARINE

THE QUEEN'S AWARD FOR EXPORT ACHIEVEMENT

| Prefix | Office of Origin | No. | No. of Words | Date filed | Time filed |
|---|---|---|---|---|---|
| Service Instructions | | Sent by | Sent to | Date Sent | Time Sent |

* DELETE IF NOT APPLICABLE

| CHARGES | | |
|---|---|---|
| Mobile Station | | |
| Land Station | | |
| Land Line | | |
| | | |
| Surcharge | | |
| | | |
| TOTAL £ | | |

| R/T | * VHF/MF/HF | Duration Mins. |
|---|---|---|

* DELETE WHICHEVER IS NOT APPLICABLE

You should be able to fill in the Preamble, including Service Instructions; Address; Text; Signature.

If you have missed any of the above, listen again as many times as necessary. Then turn to Appendix II, Unit XIII and follow the conversation in the tapescript, while you listen again. Check the information that you have added to the telegram form.

## GUIDED WRITING

### (B) Writing up a radiotelephone log

*Now listen to the instructions and the radio transmission again and write up the conversation exchange you have just listened to by completing the outline below. Write in abbreviated form, copying what has already been written on page 198 up to the end of the TR message.*

S.S. ESSEX COAST GMML
M.V.

DIARY OF THE RADIOTELEGRAPH SERVICE

| DATE AND TIME G.M.T. | STATION FROM | STATION TO | FULL DETAILS OF CALLS, SIGNALS AND DISTRESS WORKING, ETC., AS PRESCRIBED BY RULE 19 OF THE MERCHANT SHIPPING (RADIO) RULES, 1965 | FREQUENCY |
|---|---|---|---|---|
| 12.3.84 | | | | |
| 0830 | | | | 2182 |
| | | | | |
| 30/33 | | | | " |
| 35 | | | L.S.W. ON | " |
| 35 | | | | 2381 |
| 36 | | | | 1792 |
| 37 | | | | 2016 |
| 37 | | | | 1856 |
| 39 | | | | 2016 |
| 39 | | | | 1856 |
| 41 | | | | 2016 |
| 41 | | | | 1856/2016 |
| 44 | | | | 1856 |
| 45 | | | | 2016/1856 |
| 46 | | | Resume watch L.S.W. OFF | 2182 |
| | | | | |
| | | | | |
| | | | | |
| | | | | |
| | | | | |

*Note:* L.S.W. refers to Loud Speaker Watch guarding distress frequency 2182 khz. Compare your version with that given in the Answer Key, Appendix I.

# UNIT XIV

# SHIPBOARD ELECTRONICS

## READING COMPREHENSION

### (A) Description of a Marine VHF FM transmitter

In Unit XII you were asked to write a description of a basic radio communication system from notes and a diagram. Here is a simple description of a radio transmitter and receiver. *Study the description in conjunction with the diagrams.*

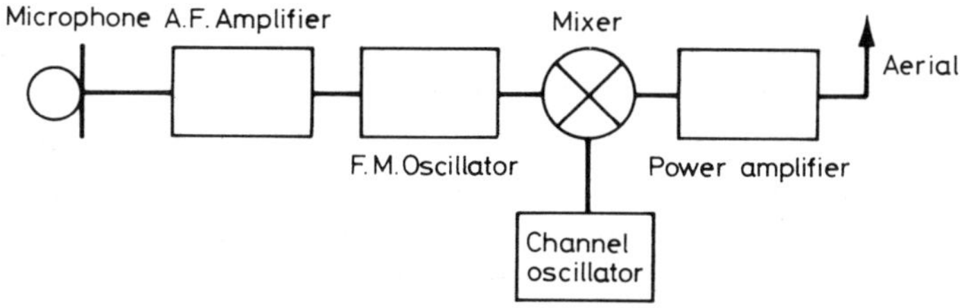


Marine VHF FM transmitter

The above diagram shows a Marine VHF FM transmitter. It provides the facility, with the associated receiver, of reliable short range communication. A typical range is about 40 miles, using the maximum transmitter power output of 25 Watts.

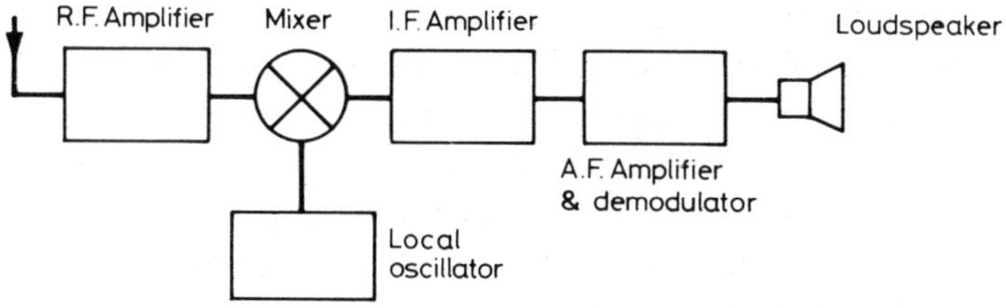


Marine VHF FM receiver

The second block diagram above shows a Marine VHF FM receiver. It is the complement of the transmitter. The total unit is known as the transceiver (**trans**mitter and re**ceiver**).

The system operates on fixed frequency channels around 160 Mhz in the VHF band. Frequency modulation is used, as this type of modulation provides clear communication without a high level of noise. The system provides for telephony only and is widely used for communication between ships, and between ships and port authorities. Provision is also made for ships to contact coastal radio stations and to be connected to the International Telephone network for the exchange of public correspondence or private telephone calls for the officers and crew members of a ship.

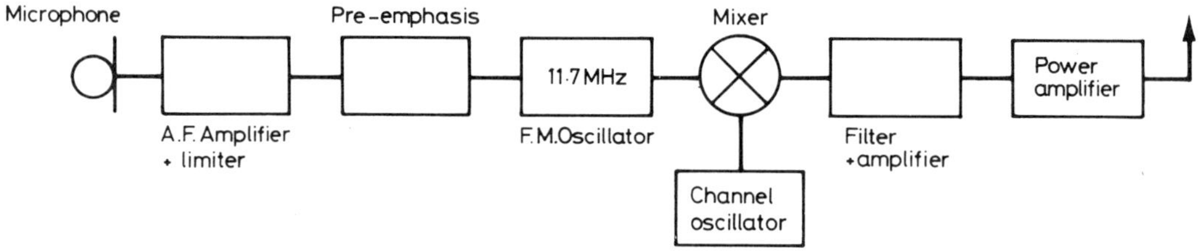


Marine VHF FM transmitter

The above block diagram shows the VHF transmitter in more detail. As the system is only for radio telephony, a microphone must be used to convert the sound energy into electrical energy. The signal is amplified in the audio frequency amplifier, but the amplitude is also limited to a pre-determined amount. This is to ensure that when frequency modulation occurs, the frequency of the unmodulated carrier wave does not vary by an excessive amount. In frequency modulation, the carrier is made to vary directly as a result of the amplitude of the audio signal. Thus, in the block diagram above a carrier signal at a radio frequency of 11.7 Mhz is frequency modulated by the audio signal. As the amplitude of the audio signal increases, the more the 11.7 Mhz carrier varies or deviates from its unmodulated value.

The pre-emphasis stage ensures that the higher audio frequencies are amplified in such a way that these components are not swamped in the noise level inherent in such systems. The frequency-modulated signal is mixed with a second carrier frequency generated in the channel oscillator. The output from the mixer contains several frequencies including the desired frequency of final transmission. Thus, the desired frequency is passed through a filter stage which rejects all the unwanted frequencies. The signal at the correct frequency is power amplified (to a maximum of 25 Watts) and the signal is radiated using the aerial. To change frequency, it is only necessary to change the channel oscillator frequency. Each frequency is allocated a channel number, for example Channel 16, which is equivalent to 156.8 Mhz.

*Now complete these notes describing the function of each unit of the FM transmitter above. Use the first, which has been done for you, as a guide.*

| | |
|---|---|
| Microphone | – Converts the sound energy into electrical energy. |
| Amplifier | – ........................................................................ |
| Limiter | – ........................................................................ |
| Pre-emphasis stage | – ........................................................................ |
| Oscillator | – ........................................................................ |
| Mixer | – ........................................................................ |
| Channel oscillator | – ........................................................................ |
| Filter | – ........................................................................ |
| Power amplifier | – ........................................................................ |

# GRAMMAR

## (A) Describing component parts, position and connection

The following verbs can be used when describing component parts, position and connection. Study their meaning and their use.

| | | |
|---|---|---|
| Component parts: | to **comprise**<br>to **consist of**<br>to **be made up of** | for specifying all the main parts of an object or system. |
| | to **contain** | for saying what is inside an object. |
| | to **include** | for giving some of the objects in a group – the list often ends with *etc.* |
| | to **be grouped** (into) | placed together. |
| Position: | **located**<br>**situated** | set or found in a particular place or position. |

| | | |
|---|---|---|
| | **mounted** (in, on) | put, usually on a base or stand (implied), and fixed ready for use. |
| | **fitted** (with) | supplied and held in place, often with bolts or screws. |
| | **installed** | put into place and connected up ready for use. |
| Connection: | **connected to/with** | joined to or related to. |

*Now study how some of the above words are used in the description below of the components of a typical marine radar installation.*

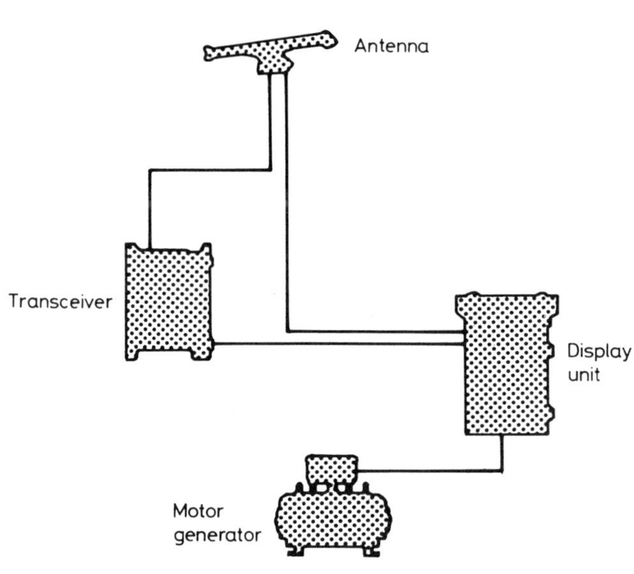


The parts of a typical radar installation **are grouped into** four main units:

1. The motor generator with its voltage regulator. This may **be mounted in** any suitable place. The motor **is connected to** the ship's mains and drives the alternator which generates the alternating voltage required for the radar installation. The voltage regulator keeps the voltage generated by the alternator constant, despite any fluctuations in the mains supply.

2. The transmitter unit and/or receiver (transceiver). This **comprises** for example, the modulator, the magnetron, part of the receiver or the complete receiver. It **is** usually **situated in** an easily accessible spot as near as possible to the aerial.

3. The indicator or display unit, which **contains** the cathode ray tube. Here the operating controls **are located**. The display unit may **be mounted in** the wheelhouse, the chartroom or, as is the trend in new ships, in a special dark room which also has facilities for plotting. The radar operator observing the display should have, as far as possible, a clear view straight ahead. If required, a second indicator unit may **be fitted** in either another room or beside the first.

*Exercise* 1. *Now choose one of the above words and put it into the appropriate form, when necessary, to complete the description below.*

4. The aerial unit or antenna assembly, which ............. the reflector, aerial driving motor, synchro-generator, etc. The transmitter and early stages of the receiver are sometimes ............. in the aerial unit. With this arrangement the problem arises of keeping the components dry and repairing them, especially in wet weather. The presence of water or moisture in the waveguides, etc., reduces the radiation materially, if not completely and it is therefore common for electrical heating equipment to be ............. . A thermostat may sometimes be ............. to keep the temperature of the apparatus at a constant level. To facilitate the repair or inspection of the aerial, a telephone ............. to the indicator unit is sometimes ............. near the aerial. A switch may also be ............. there for switching on and off the driving motor of the aerial. The scanner should, as far as possible, be ............. well clear of the obstacles almost always present on shipboard, e.g. funnels, derricks, ventilators, etc., to give an uninterrupted sweep of the horizon. It should be ............. as high as possible to increase the radar's range; but as the angle of incidence increases, the sea echoes become more pronounced, and this is not desirable.

## (B) Instructions

Instructions in English are usually given by using the **imperative**. This is formed by the infinitive without 'to'.

e.g. Turn on the mains.
Select the channel.

Sometimes the instruction is based on a condition.

e.g. If the mains lamp lights, select and key transmitter.
If the mains lamp does not light, check the mains is connected.

*Study this fault finding chart.*

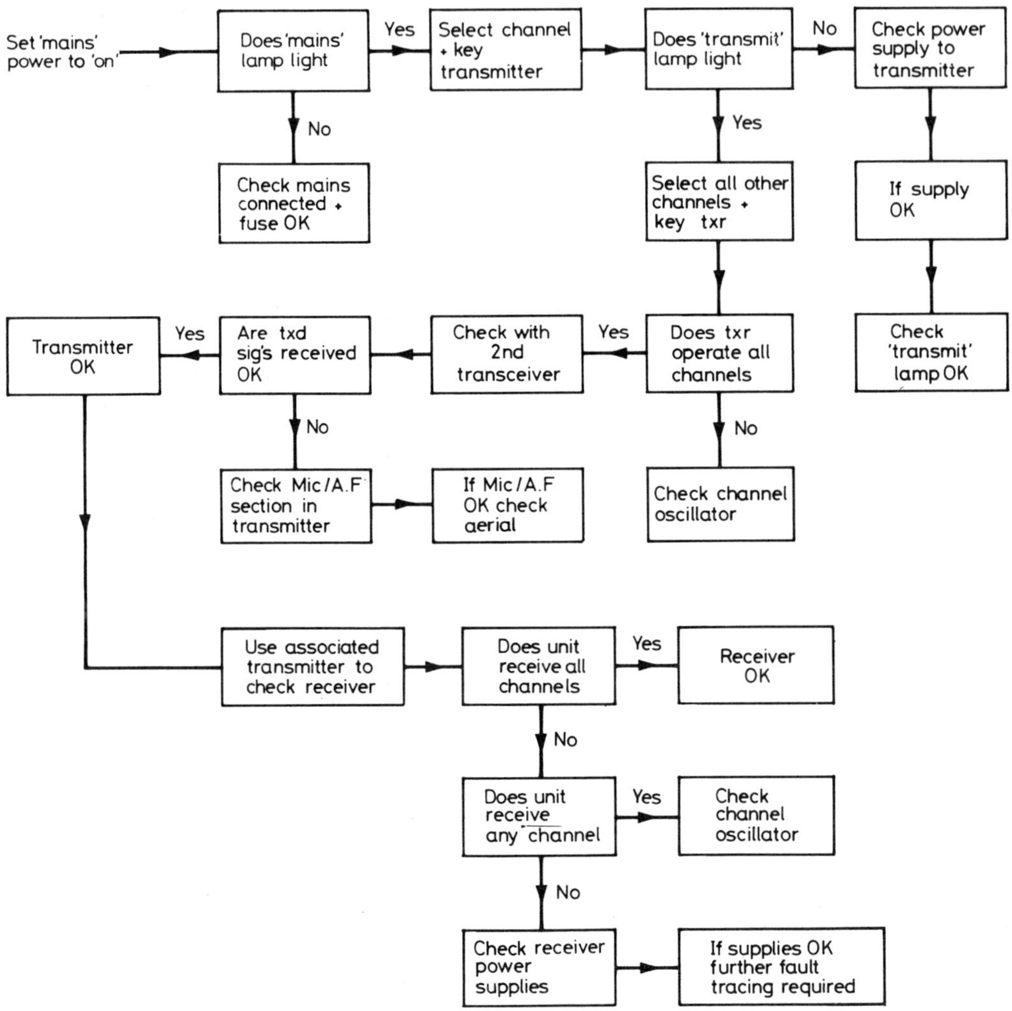


Fault Finding Chart for Marine VHF Transmitter/Receiver

*Key:* Txr = Transmitter; Txd = Transmitted; Mic = Microphone; A/F = Audio Frequency.

*Exercise* 1. *Use the chart to complete the instructions below for checking a transmitter.*

1. Set the mains power to 'On'.

2. If the mains lamp does not light, ........................................

3. .................................... , select channel and key transmitter.

4. If the transmit lamp does not light, ......................................

5. ...................... , select all other channels and key transmitter.

6. If the transmitter does not operate on all channels, ................

.......... .

7. ............................................ , check with 2nd transceiver.

8. If the transmitted signals are not received OK, .......................

.............. .

9. .................................................. , the transmitter is O.K.

*Exercise* 2. *Look at the fault chart again. What would you do in the following circumstances.*

(a) The transmit lamp does not light, although the power supply is OK.
(b) Transmitted signals are not being received, but the Mic/AF. section is OK.
(c) You need to check the receiver. What would you use?
(d) The receiver only receives some channels.
(e) The receiver does not receive any channels.

# READING COMPREHENSION

## (B) The echo sounder

*Read through the following description of echo sounders and do the exercises as you come to them.*

The echo sounder is an electrical device for measuring the depth of the water. It is now installed in almost every ship.

The principle of the measurement is as follows: the echo sounder sends a radio signal from below the ship to the seabed, where it is reflected back. The time taken to receive the reflected signal is a measure of the depth of the water under the ship. The radio pulse may be as short as 1 millisecond (thousandth of a second) and the frequency between 10 and 50 kiloHertz (thousands of cycles a second). The received pulse is displayed on a chart by a pen recorder so that the navigator can see the outline of the bottom over which the vessel is passing.

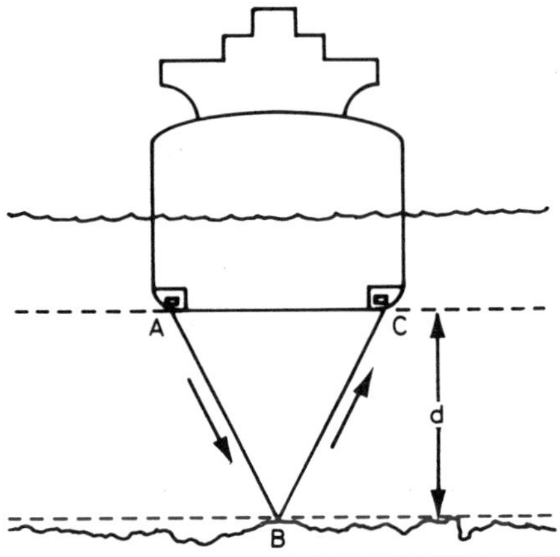


Fig. 14.1 Principle of the echo sounder. A short pulse of sound vibration is produced at $A$, reflected by the seabed at $B$, and received at $C$. The depth $d$ is proportional to the measured time interval between the transmission and reception

If the water is not too shallow, the 'Pythagoras error', due to the distance $d$ being less than $AB$, may be ignored.

The velocity of propagation of sound pulses in seawater is practically constant and generally taken as 1500 metres per second. Therefore it follows that:

$$2d = vt \quad \text{or} \quad d = \frac{vt}{2}$$

where $v$ = velocity of sound in water and $t$ = time interval.

*If the time interval between transmission of the pulse and reception of the echo is one second, what is the depth of water under the ship?*

*If the time interval is 0.5 seconds, what is the depth?*

In shallow water it is dangerous to neglect the Pythagoras error, because the distance AB will be greater than the true depth. Although depth is shown on charts, the seabed, particularly near coastlines, can change considerably. It is not always possible to indicate these changes immediately on the charts.

An echo sounding apparatus consists of five basic components and operates as follows. Ultra-sonic oscillations are generated in the oscillation generator. These are of sufficient power to return echoes from several hundred fathoms. They are then supplied to the transmitting oscillator, which may be of the Piezo-electric or magnetostrictive type. The transmitting oscillator starts vibrating when the electrical oscillations are supplied to it, while the receiving oscillator, which is set vibrating by the echo, converts the mechanical vibration back into an electrical oscillation. The amplifier is used for strengthening the weak oscillations which the receiving oscillator has converted from sound vibrations. These oscillators can also be called transducers. The indicator or recorder measures and indicates the depth. Echo sounders can be divided into two kinds: recorders (echographs) and echometres. Echometres only indicate instantaneous depth, whereas echographs also record depths. If a vessel moves at a constant speed, you can obtain an automatically recorded profile of the bottom of the sea along the route of the ship.

*Use the description above to help you label Fig. 14.2 showing the components of an echo sounding apparatus. The first has been done for you.*

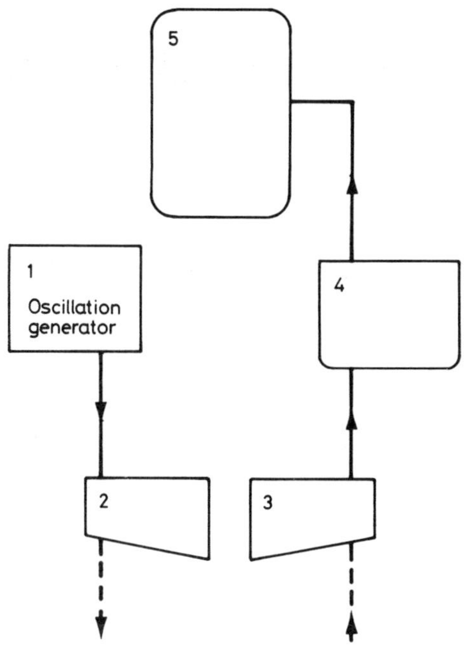


Fig. 14.2 The components of an echo sounding apparatus

The true depth of the water is equal to the sum of the depth indicated by the echo sounder *d* in Fig. 14.1, and the vertical distance between sea level and the transducers. The echo sounder can be checked by taking a sounding with the hand leadline. This can be done while waiting for a pilot to be taken on board. The seabed should be flat or slope gently. Two or more soundings are then taken by lead as close as possible to the transducer.

The number of transducers varies. One only may be fitted, used for both transmission and reception; however, there are often two, both connected to the same indicator, one serving for transmission and the other for reception. A transducer can be connected to two recorders or to an echometre and a recorder. In this case the echometre is installed on the bridge for instantaneous reading of the depth, while the recorder is installed in the chart room. Large ships may be equipped with two transducers: one fore and one aft, each operating as both transmitter and receiver. Very large ships (VLCCs and LNGCs) may have either fore and aft dual transducers, one on the port side and one on the starboard side, or they may have midships dual transducers, one on the port side and one on the starboard side, and fore and aft single transducers.

*Draw two sketches of a ship's hull to show the two possible configurations for the arrangement of transducers in very large ships.*

# APPLIED TERMINOLOGY

## (A) Terms relating to circuit diagrams

*Study these symbols which are in common use on circuit diagrams in manuals.*

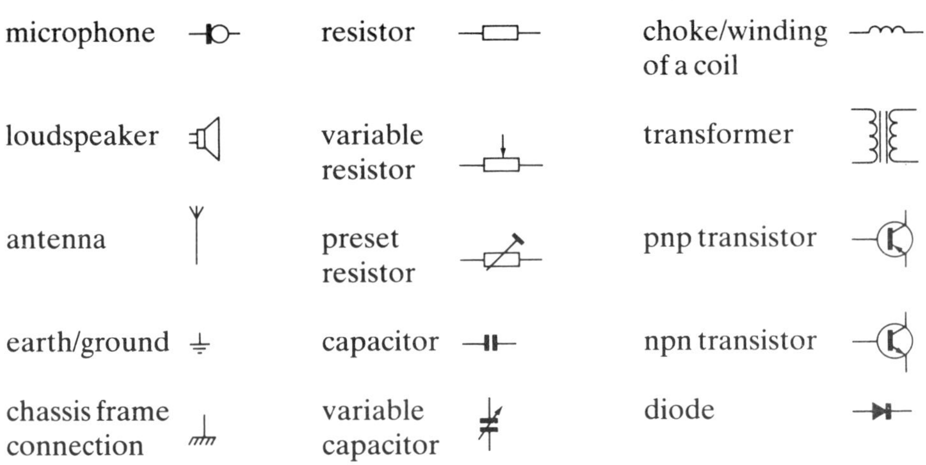

| | | |
|---|---|---|
| switch | preset capacitor | Zener diode |
| fuse | | varactor |
| cell | electrolitic capacitor | crystal |

*Exercise* 1. *Now study this circuit diagram of an oscillator and amplifier. Write in the names of the components 1-8 against the appropriate number below.*

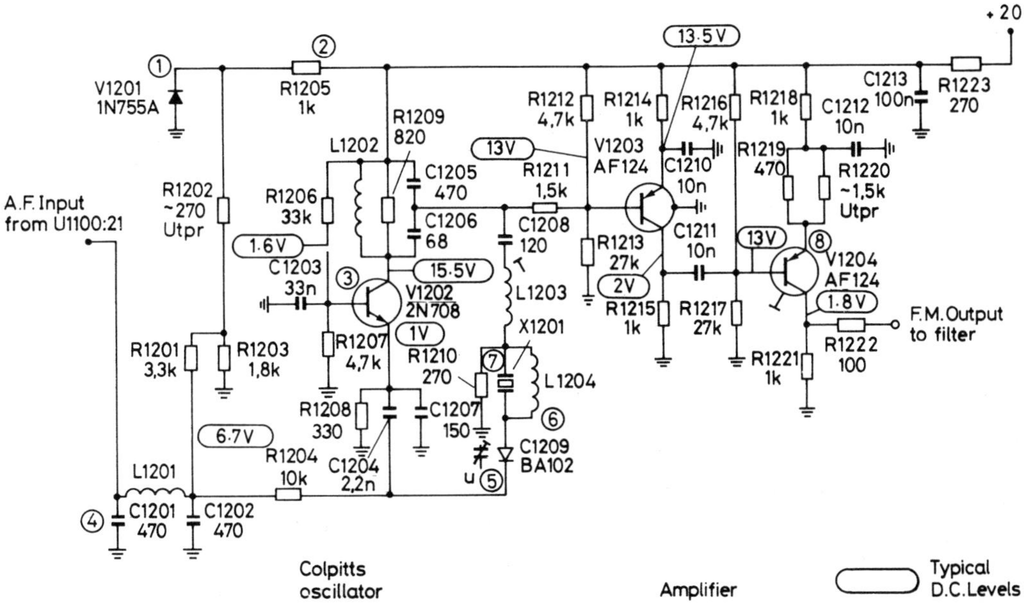


1 ..................... 2 ..................... 3 ..................... 4 .....................

5 ..................... 6 ..................... 7 ..................... 8 .....................

*Note:* This circuit is that belonging to the Frequency Modulated Oscillator, U1200 on page 221.

# (B) Measurement

## (xiv) ac waveforms

*Read through the following description of sinusoidal ac waveform and do the exercises as you come to them.*

Many waveforms are possible with alternating currents (ac). One of the simplest to produce comes by rotating a loop of wire in a uniform magnetic field. It is called a **sinusoidal** waveform and is shown in Fig. 14.3.

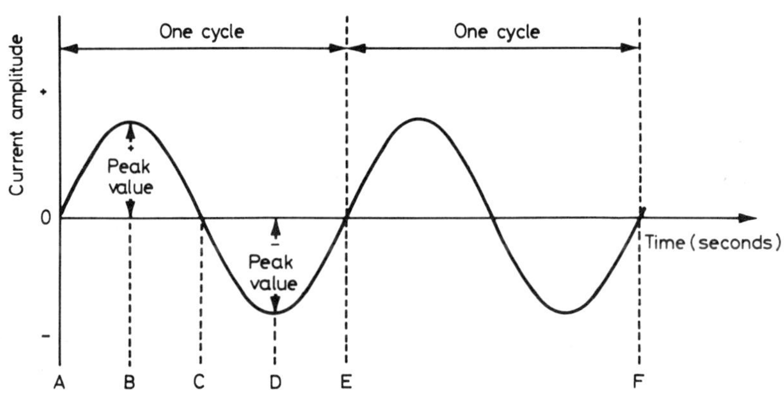


Fig. 14.3

Between points *A* and *B* the current increases from zero to a peak value in the positive (+) direction. Between points *B* and *C* the current gradually reduces to zero. Then, between points *C* and *E* the same occurs as for *A* and *C*, but in the opposite or negative (−) direction. The whole sequence from *A* to *E* represents one complete rotation of the wire loop in the magnetic field, and is called one **cycle** of ac waveform.

The time needed, in seconds, for one cycle of waveform to be produced is called the **periodic time** (*T*). The number of complete cycles occuring in one second is called the **frequency** (*f*) and is measured in **Hertz** (Hz). One Hertz is one cycle per second.

*Exercise* 1. *Study Figure 14.4.*

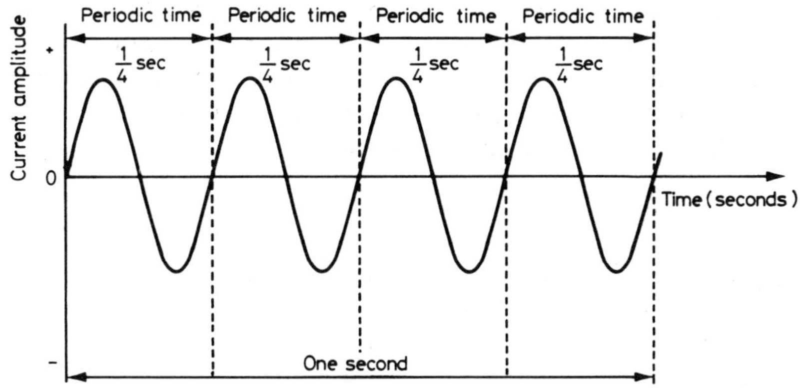


Fig. 14.4

If $f = \frac{1}{T}$ and $T = \frac{1}{f}$

(a) *What is the frequency of the waveform in Fig. 14.4?*
(b) *What is the periodic time of the waveform in Fig. 14.4?*

The maximum amount by which a waveform deviates from its mean value is called the **amplitude** of the waveform.

Because the sinusoidal waveform is produced by rotating a loop of wire in a magnetic field, we could consider the loop as moving through 360° in one rotation. Thus, we could plot the resultant current against angular rotation, as shown in Fig. 14.5.

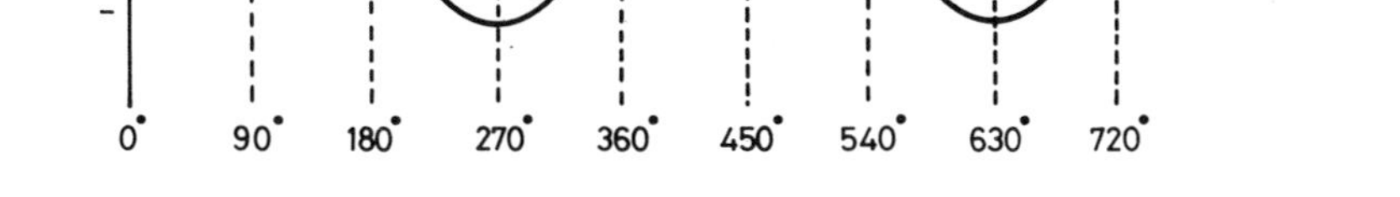


Fig. 14.5

Also, the energy of an ac waveform travels through space or along a transmission line at a particular velocity, therefore a certain distance is travelled in the periodic time for one cycle. In Fig. 14.6, we can see that the ac waveform repeats complete cycles over distances. The distance representing each cycle is called the **wavelength** (λ) and is measured in metres.

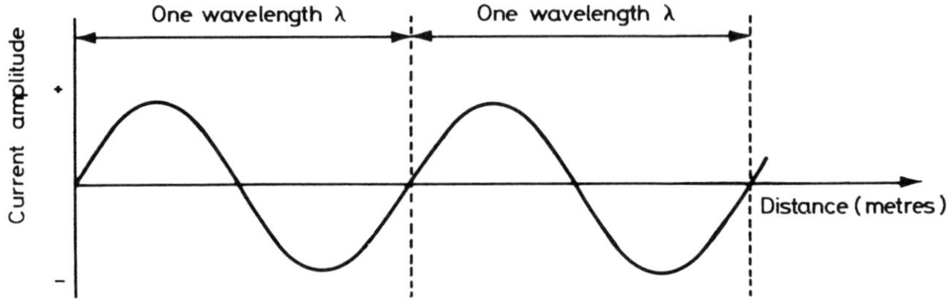


Fig. 14.6

The timing of a waveform in relation to a reference point or another waveform is known as the **phase** of the waveform. If two waveforms are identical except for their phase, the difference between the two can be expressed as a **phase difference**. In Fig. 14.7, waveform *A* is seen to be leading waveform *B* by 90°, or waveform *B* is lagging waveform *A* by 90°.

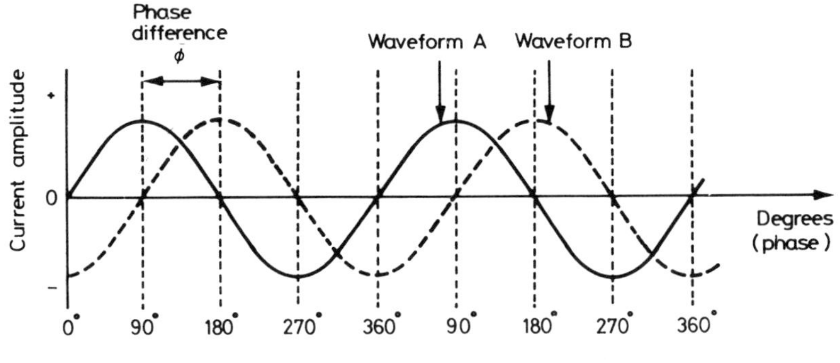


Fig. 14.7

*Exercise 2. Complete the table overleaf to show the relationship between frequency, wavelength and velocity.*

| *Formula* | *Meaning in full* |
|---|---|
| $f=\frac{v}{\lambda}$ | .................................................... |
| .......... | Wavelength equals velocity divided by frequency. |
| $v=\lambda f$ | .................................................... |

# GUIDED WRITING

## (A) Description of a Marine VHF FM receiver

*Use the diagram and the notes below to help you write a simple description of a Marine VHF FM receiver. Begin with a suitable introductory sentence relating your description to the diagram.*

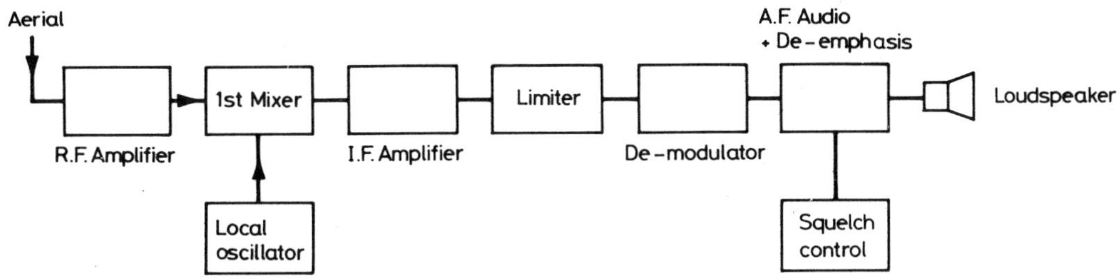


Aerial – Intercepts the modulated radio frequency carrier signals.

RF Amplifier – Signal selection takes place; the signal is amplified.

1st. mixer
Local oscillator – The amplified r.f. signal is coupled to the input of a mixer stage; it is combined with the output of a local oscillator. The two frequencies beat together in the mixer, they generate an intermediate frequency.
This frequency is equal to the difference between the rf signal and local oscillator frequency.
The intermediate-frequency (i.f.) signal at the output of the mixer is lower in frequency than the original r.f. carrier signal; it contains the same modulation as the original r.f. carrier signal.

IF amplifier – The i.f. signal from the mixer is amplified by several stages of fixed-tuned i.f. amplification.

| | |
|---|---|
| Limiter | – Feeds the demodulator with a constant amplitude signal removes any AM component and peaks of noise which may be present. |
| Demodulator | – Takes the varying frequency signal and converts it to varying amplitude audio signal. |
| AF amplifier De-emphasis | – The audio-signal from the demodulator is amplified by one or more stages of audio amplification until it is strong enough to drive a loudspeaker. |
| Squelch control | – Mutes af stage until the signal is present at the receiver; without this unit the intrusive squelch noise would be present all the time. The signal switches out the squelch action. |
| Loudspeaker | – Converts the audio signal into sound waves corresponding to the original sound of the message. |

# NOTE-TAKING PRACTICE

## **Lecture:** A VHF transceiver

*Study this block diagram of a Marconi Argonaut S transceiver. The simplified diagrams in Reading Comprehension (A) and Guided Writing (A) should help you to identify some of the stages in each of the units in the transmitter and receiver.*

*Now listen to this description of what happens to a signal as it passes through each unit of the transmitter and receiver. Follow the path of the signal on the diagram and take down detailed notes on a separate sheet of paper. Also take down notes about the Control Unit and the Power Supply.*

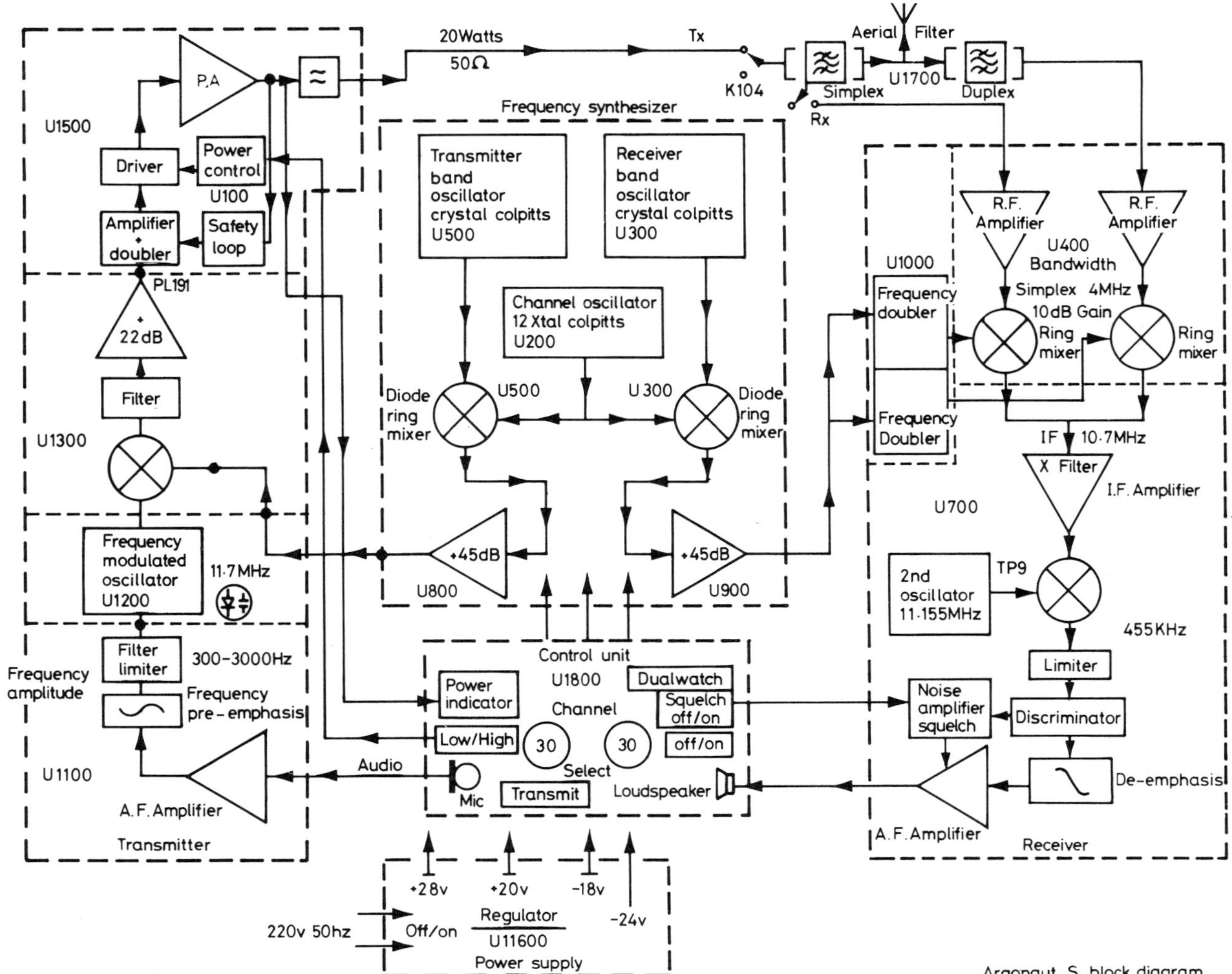


Argonaut S block diagram

# GUIDED WRITING

## (B) Description of an Argonaut S transceiver

*Write a description of an Argonaut S transceiver, using the block diagram and your notes to help you. Do not forget to write a short introductory paragraph.*

*When you have finished, compare your description with the tapescript of the lecture in Appendix II.*

# APPENDIX I

# KEY TO EXERCISES

## UNIT I SHIPPING

### Reading comprehension

#### *(A) How merchant ships operate (version 1)*

The three basic ways merchant vessels can operate:

(a) as liners
(b) as tramps
(c) as specialized vessels

(a) Merchant ships are designed to carry cargo and passengers.
(b) Liners are employed on regular routes on a fixed timetable.
(c) Tramps are not employed on regular routes and do not keep to a fixed timetable.
(d) Specialized vessels are designed to carry a particular type of cargo.

### Grammar

#### *(A) Quantifiers*

*Exercise* 1.

| Quantifier | Countable noun | Uncountable noun |
|---|---|---|
| several | passengers | – |
| no | passengers | money |
| a lot of | passengers | money |
| few | passengers | – |
| much | – | money |
| a large number of | passengers | – |
| a little | – | money |
| all | passengers | money |
| many | passengers | – |
| a large amount of | – | money |
| a few | passengers | – |
| some | passengers | money |
| most | passengers | money |
| little | – | money |

*Exercise* 2.

A few passengers travel by cargo liner.
A large amount of money is needed to operate a shipping fleet.
Few passengers enjoy being at sea when it is rough.
No passengers are allowed down in the engine room.
All money on board ship should be kept in a safe place.
(*Note:* There are other possible answers.)

### *(B) Logical connectives*

*Exercise* 1.

(a) A cadet can train as a Deck Officer or an Engineer Officer.
(b) The ship was old, but she was in good condition.
(c) Passenger liners carry passengers and some cargo.
(d) Tankers usually operate as specialized vessels, but they sometimes operate as tramps.
(e) Merchant ships are designed to carry liquid or dry cargo.
(f) Warships are designed for speed and manoeuvrability.

*Exercise* 2.

(a) He failed both the practical exam and the written exam.
He failed not only the practical exam but also the written exam.
(b) A Cadet can train as either a Deck or an Engineer Officer.
(c) The crew saved both the ship and the cargo.
The crew saved not only the ship but also the cargo.
(d) They had neither food nor water for two days.
(e) Some ferries carry both passengers and vehicles.
Some ferries carry not only passengers but also vehicles.
(f) Merchant ships operate as either tramps or liners.

## Reading comprehension

### *(B) How merchant ships operate (version 2)*

| | |
|---|---|
| [1]Merchant ships | ← [2]some |
| [1]Merchant ships | ← [3]a few |
| [4]Merchant vessels | ← [5]They |
| [5]liners | ← [6]These |
| [5]liners | ← [7]their |
| [5]liners | ← [7]they |
| [8]deep-sea liners | ← [9]The former |
| [8]short-sea liners | ← [9]the latter |
| [10]Ferries | ← [11]These |
| [12]passenger liners | ← [13]They |
| [15]tramps | ← [16]These vessels |

| | |
|---|---|
| [15]tramps | ←[16]them |
| [17]Tramps | ←[18]A number |
| [18]coasters | ←[19]These |
| [17]Tramps | ←[20]some |
| [21]specialized vessels | ←[22]These |
| [21]specialized vessels | ←[24]The most common |
| [24]oil tankers | ←[25]They |

(a) liners – deep-sea liners (containerized)
short-sea liners (containerized or conventional)
ferries (passengers and vehicles)
passenger liners (passengers and some cargo)

(b) tramps – deep-sea tramps (dry bulk, some general)
short-sea tramps (dry bulk, some general)
coasters (dry bulk, some general)

(c) specialized vessels – oil tankers (liquid bulk)
chemical carriers (liquid bulk)
LNG carriers (liquid bulk)

(a) Most passengers now travel by air because it is quicker and often cheaper.

(b) There are not enough passengers at certain times of the year (e.g. during winter months) to make it economical for them to operate on the liner routes during these periods.

(c) Most tramp cargoes are dry bulk cargoes.

(d) Other specialized vessels include: fruit carriers, timber ships, refrigerated vessels, container ships.

## Applied terminology

### *(A) Terms relating to shapes*

*Exercise* 1.

(a) It is shaped like a cone. It is conical in shape.

(b) It is shaped like a cylinder. It is cylindrical in shape.

(c) It is shaped like a triangle. It is triangular in shape.

(d) It is shaped like an ellipse/oval. It is elliptical/oval in shape.

(e) It is shaped like a sphere. It is spherical in shape.

(f) It is shaped like a semicircle. It is semicircular in shape.

(g) It is shaped like a circle. It is circular in shape.

(h) It is shaped like a square. It is square in shape.

(i) It is shaped like a rectangle. It is rectangular in shape.

*Exercise* 2.

(a) a T-beam
(b) an eye splice
(c) an A-bracket
(d) a heart shackle
(e) an L-shaped cabin
(f) a mushroom ventilator
(g) a needle valve
(h) a D-shackle
(i) finger plates
(j) a V-shaped cut
(k) a kidney tray
(l) a U-shaped tube.

## (B) Measurement

*Exercise* 1.

| Noun | Adjective |
|---|---|
| width | wide |
| breadth | broad |
| height | high |
| depth | deep |
| thickness | thick |

# Guided writing

## (A) Description of buoys

*Stage* 2.

Starboard hand buoys are green in colour. They are either cone-shaped, pillar-shaped or spar-shaped. If they carry a top mark it is also green and shaped like a cone. Lights when fitted, are green and flash in any rhythm.

*Stage* 3.

Special marks are yellow in colour. They are either can-shaped, sphere-shaped, cone-shaped, pillar-shaped or spar-shaped. If they carry a top mark, it is also yellow and shaped like a cross. Lights, when fitted, are yellow and flash in any rhythm.

# UNIT II SHIP TYPES

## Reading comprehension

### *(A) Types of merchant ships (version 1)*

Merchant ship types

| Cargo ships | | Passenger ships |
|---|---|---|
| Dry cargo ships | Liquid cargo ships | |
| Multi-deck vessels | Oil tankers | Passenger liners |
| Bulk carriers | LNG carriers | Cruise ships |
| Container ships | Chemical carriers | Ferries |
| Refrigerated ships | | |

Merchant ships can be designed as cargo ships or passenger ships. Cargo ships can be divided into dry cargo ships and liquid cargo ships. Dry cargo ships include multi-deck vessels, bulk carriers, container ships and refrigerated ships. Oil tankers, LNG carriers, and chemical carriers are examples of liquid cargo carriers. Three types of passenger ship are passenger liners, cruise ships and ferries.

## Grammar

### (A) Articles

*Exercise* 1.

The largest type of cargo ship is the tanker. Tankers are designed to carry liquid cargo such as oil. The cargo is pumped directly into holds by powerful pumps. The holds are constructed as tanks. The tanks are subdivided into a central tank, two wing tanks and an expansion tank. The expansion tank allows the oil to expand in hot weather. The bridge superstructure and the engine room are situated aft to leave more room for cargo. The bridge is connected to the forecastle by the catwalk. Tankers which are over 500,000 dwts are known as ultra large crude carriers (ULCCs).

*Exercise* 2.

(a) Ships of all types have crossed the Atlantic.
(b) The largest ships have crossed the Atlantic.

(c) Merchant ships have crossed the Atlantic.
(d) The ships which arrived today have crossed the Atlantic.
(e) The ships loading in the port have crossed the Atlantic.
(f) Most ships have crossed the Atlantic.

## (B) Logical connectives

*Exercise* 1.

(a) Multi-deck vessels usually carry general cargo; however, some carry containers as well.
(b) Passenger liners have high superstructures because they need a large number of decks.
(c) Many ferries are designed to carry vehicles, therefore they have doors at the bows or stern.
(d) Cargo ships are usually designed to carry dry or liquid cargo; however, OBO (oil, bulk ore) ships are designed to carry both.
(e) Bulk carriers carry large quantities of loose cargo, therefore they have large unobstructed holds.
(f) Passenger liners often operate as cruise ships for part of the year because there is not always enough business for them on liner routes.

*Exercise* 2.

(a) Although multi-deck vessels usually carry general cargo, some carry containers as well.
(b) Passenger liners have high superstructures as they need a large number of decks.
(c) Many ferries are designed to carry vehicles, consequently they have doors at the bows or stern.
(d) Although cargo ships are usually designed to carry either dry or liquid cargo, OBO (oil, bulk ore) ships are designed to carry both.
(e) Bulk carriers carry large quantities of loose cargo, consequently they have large unobstructed holds.
(f) Passenger liners often operate as cruise ships for part of the year as there is not always enough business for them on liner routes.

# Reading comprehension

## (B) Types of merchant ships (version II)

| | |
|---|---|
| [1]Merchant ships | ← [1]they |
| [1]Merchant ships | ← [2]Most |
| [1]Merchant ships | ← [2]a few |
| [3]Cargo ships | ← [4]One type |
| [3]Cargo ships | ← [4]the other |

| | | |
|---|---|---|
| [4]dry cargo . . . liquid cargo | ← | [4]both |
| [5]the multi-deck vessel | ← | [6]Her |
| [6]'tween decks | ← | [6]these |
| [7]bulk carriers | ← | [8]These |
| [9]the container ship | ← | [10]They |
| [12]Oil tankers | ← | [13]They |
| [12]Oil tankers | ← | [13]ones |
| [16]the passenger liner | ← | [16]many |
| [16]the passenger liner | ← | [17]their |
| [18]the cruise ship | ← | [19]These |
| [20]the ferry | ← | [21]Many of them |
| [20]the ferry . . . [21]designed to carry vehicles | ← | [21]these |

(a) . . . ; however, an OBO ship can carry both.
(b) . . . , because these make stowage of individual packages easier.
(c) . . . , consequently stowage is easier.
(d) . . . , because huge quantities of oil need to be transported, and one large vessel is more economical to operate than two smaller ones.
(e) . . . ; however, they can also be carried in drums in general cargo ships.
(f) . . . ; however, many carry cargo as well.
(g) . . . , therefore they have doors at the stern or bows.
(Some variation to the answers given above is possible.)

(a) It increases the type of cargo the ship can be used to carry and the chances of the ship getting cargo for the return journey.
(b) They act as shelves and make it easier to stow heavy cargo which must be unloaded first.
(c) One large tanker which can carry the same amount of oil as two smaller ones costs less for fuel, maintenance and crew than the cost for the two smaller ones.
(d) Chemicals are not necessarily needed in large quantities and chemical carriers do not operate on all routes.
(e) It enables all available space to be put to economic use.
(Other explanations are also possible in addition to those given above.)

## Guided writing

### *(A) Description of different types of warship*

In the past the largest warships were battleships. They were designed for heavy bombardment, therefore they carried up to sixteen-inch guns. Their speed was relatively slow because they were large and had heavy armour plating.

Nowadays aircraft carriers are the largest warships. Although most carry aircraft and helicopters, a few carry helicopters only. They have a narrow bridge superstructure because this leaves more room for the flight deck.

Another type of warship is the cruiser. These are smaller than battleships, but larger than destroyers. They are designed to combine fire-power with speed, therefore they have medium-size guns and missiles.

Destroyers and frigates are designed for escort duties. The modern destroyer is taking over the role of the cruiser and it carries guided missiles in addition to conventional guns. Many frigates have an anti-submarine role, therefore they carry weapons for destroying submarines.

Submarines ............................................................................................

............................................................................................................

Patrol boats vary greatly in design. They are built for speed and manoeuvrability. They are often powered by gas turbines, but these are expensive to run, therefore they may have diesel engines for cruising.

(Other connecting words may be used to join these sentences. The paragraphs are arranged in order of size of ship, the largest (battleships) being first.)

# UNIT III SHIP CONSTRUCTION

## Reading comprehension

### *(A) Building ships*

(a) For example, a general cargo vessel costs several million pounds and a giant tanker can cost over £40 million pounds.
(b) This is because of the high cost of steel and other materials used in shipbuilding and the high cost of labour.
(c) With the result that many of the old processes have disappeared or been combined into one fully mechanized process, and machines are now used instead of men.
(d) Consequently, ships can now be built in about sixteen months and costs can be kept to a minimum.
(e) The largest shipping companies, for instance, have their own naval architects.
(f) In Europe and Japan, shipyards employ naval architects to design a ship for a customer, or alternatively, offer basic designs which can be varied to suit the customer's needs.
(g) In addition, shipowners may also go to independent firms of shipping consultants and ask their naval architects to design ships for them.
(h) When shipowners decide to order a new ship, they not only tell the naval architect the cargo they want the ship to carry, but also the routes the ship will ply and the desired speed.

(i) In addition, they put limits on the ship's dimensions and on the price that they are prepared to pay.
(j) Furthermore, the ship must comply with the rules of the classification society and with international regulations.
(k) To sum up, economic, engineering and safety factors all govern the design of a ship.

## Grammar

### (A) Passives

*Exercise* 1.

(a) Nowadays ships are made of steel.
(b) The tanker will be launched by the Queen tomorrow.
(c) The hull is being painted.
(d) Two new warships have been ordered by the Royal Navy.
(e) Ships used to be built of wood.
(f) Two pieces of metal can be joined together by welding.
(g) Next, the piece of metal was weighed.
(h) The plates must be prepared properly.
(i) The bows had to be repaired.

*Exercise* 2.

When a ship is fitted out, she is completed. The engines are put in, if they have not already been installed. The superstructure is finished off and the accommodation for the crew is constructed. In addition, masts and derricks are erected and various items of deck machinery are put in place. Outside companies usually do the electrical work, plumbing and any woodwork. The furniture and fittings for all the saloons and cabins must also be bought. All this work which is done after the vessel has been launched, is called 'fitting out'.

### (B) Time relaters

*Exercise* 1.

First, the plans are completed by the naval architects. Then, they are approved by the classification society. After that, the parts of the ship are prepared. Afterwards, they are put together. Later, the ship is launched. Then, the ship is fitted out and completed. Eventually, the ship goes for sea trials. Finally, the ship is handed over to her new owners.

*Exercise* 2.

(a) After the parts of the ship are prepared, they are put together.

(b) After the ship is launched, she is fitted out and completed.
(c) Before the ship is handed over to her new owners, she goes for sea trials.

*Exercise* 3.

(a) After being prepared, the parts of the ship are put together.
(b) After being launched, the ship is fitted out and completed.
(c) Before being handed over to her new owners, the ship goes for sea trials.

## Reading comprehension

### *(B) Building ships (cont.)*

| | | |
|---|---|---|
| [2]the plans are completed | ↔ | [3]the final plans |
| [3]classification society | ↔ | [4]classed |
| [3]classification society | ↔ | [5]society |
| [4]classed | ↔ | [6]classification |
| [7]chain of shops | ↔ | [8]Each shop |
| [8]Each shop | ↔ | [9]preparation shop |
| [12]cut | ↔ | [13]Cutting |
| [12]shaped | ↔ | [13]shaping |
| [9]plates and bars | ↔ | [14]pieces |
| [8]Each shop | ↔ | [14]prefabrication sheds |
| [15]riveting | ↔ | [16]Riveting |
| [14]welded | ↔ | [15]Welding |
| [14]section | ↔ | [18]The prefabricated sections |
| [21]slipway | ↔ | [21]slide |
| [22]dry dock | ↔ | [23]The dock |
| [24]completed | ↔ | [25]A completed ship |

(1) The order for the ship is placed. (e)
(2) The final plans are drawn up. (h)
(3) The plans receive the approval of the classification society. (c)
(4) The parts of the ship are prepared. (f)
(5) The parts are welded together to form sections. (i)
(6) The sections are assembled on the building berth. (g)
(7) The ship is launched. (a)
(8) The ship is fitted out. (j)
(9) The ship and her equipment are tested at sea. (b)
(10) The new owners take possession of their vessel. (d)

## Applied terminology

*Exercise* 1.

| | |
|---|---|
| compartments | |
| bulkheads | fore peak |
| after peak | engine room |
| holds | 'tween decks |
| superstructure | forecastle |
| bridge | poop |

*Exercise* 2.

The diagram above shows the layout of a modern tanker. The hull is divided up into a number of watertight compartments by decks and steel bulkheads. At the fore and after ends of the hull are the fore peak tanks and the after peak tanks. The engine room is situated at the after end of the ship to leave more room for cargo. Between the engine room and the cargo space is a coffer dam. The cargo space is divided up into a number of tanks. Above the main deck is the superstructure. At the fore end is the forecastle. At the after end the bridge superstructure and the poop are combined.

### *(B) Measurement*

*Exercise* 1.

(a) Displacement
(b) Standard displacement
(c) Load displacement
(d) Light displacement.

## Guided writing

### *(A) Description of building a ship's hull*

(e); (k); (c); (l); (h); (g); (b); (j); (i); (d); (a); (f).

First the steel plates and bars are off-loaded from lorries or railway wagons. Then, they are stored in the stockyard ready for processing. Eventually, they are conveyed to the preparation shop by magnetic cranes. After being cleaned by shot blasting, they are coated with primer paint to prevent corrosion. After that, they are transferred to the platers' shop by conveyor rollers. Before being shaped by different presses and rollers, they are cut to the correct size by gas torches. Next, they are transferred to the assembly shop by cranes or conveyor rollers. In the assembly shop they are joined together to form larger units. Later, the units are carried to the building berth by giant cranes. Finally, at the building berth the units are welded together to form the hull.

# UNIT IV MANNING

## Reading comprehension

### *(A) The organization of a ship's crew*

Master

| **Deck Department** | **Engine Department** | **Catering Department** | | **Radio Department** |
|---|---|---|---|---|
| Chief Officer<br>Second Officer<br>Third Officer<br>(Fourth Officer)<br>(Cadets) | Chief Engineer<br>Second Engineer<br>Third Engineer<br>Fourth Engineer<br>(Fifth Engineer)<br>(Electrical Officer)<br>(Cadets) | Chief Steward | | Chief Radio Officer<br>(Second Radio Officer)<br>(Third Radio Officer) |
| | | **Galley Section** | **Saloon Section** | |
| Boatswain<br>Carpenter | Storekeeper<br>Donkeyman<br>Pumpman (Tkrs) | Ship's Cook | Second Steward | |
| AB<br>EDH<br>OS | Firemen<br>Greasers | Cooks<br>Junior Ratings | Stewards<br>Junior Ratings | |

## Grammar

### *(A) Function*

*Exercise* 1.

(a) The Chief Officer is responsible to the Master for the Deck Department.
(b) The Third Officer is responsible for the life-saving equipment.
(c) The sounding of tanks and bilges is the responsibility of the Carpenter.
(d) The Radio Officer is responsible for radio communications.
(e) The Chief Steward is responsible to the Master for the Catering Department.

(f) The preparation of food is the responsibility of the Ship's Cook.
(g) The Chief Engineer is responsible for the efficient running of his department.
(h) The loading and unloading of oil is the responsibility of the Pumpman.

*Exercise* 2.

(a) A thermometer is used for measuring temperature.
A thermometer measures temperature.
We measure temperature with a thermometer.

(b) A fire extinguisher puts out fires.
We put out fires with a fire extinguisher.
The function of a fire extinguisher is to put out fires.

(c) We raise and lower the anchors with a windlass.
The function of a windlass is to raise and lower the anchors.
A windlass is used for raising and lowering the anchors.

(d) The function of a chronometer is to measure time.
A chronometer is used for measuring time.
A chronometer measures time.

### *(B) Time relaters*

*Exercise* 1.

(a) When (b) While (c) while
(d) When (e) While (f) when

*Exercise* 2.

(a) When undergoing speed trials, the ship produces her maximum speed.
(b) Cannot be reduced.
(c) While running/being run, the engines are carefully checked.
(d) When inspecting the lifeboats, the surveyor examines their equipment as well.
(e) Cannot be reduced.
(f) While covering the mile distance, the ship keeps a straight course.

## Reading comprehension

### *(B) The deck department*

Here are some meaning links:
[2]course ↔ [3]on course
[1]navigating ↔ [2]navigator ↔ [3]navigation
[4]the cargo is stowed ↔ [5]The stowage of cargo

[5]Chief Officer ↔ [7]First Mate
[9]maintaining ↔ [10]cleaning, painting and repair work ↔ [12]maintenance
[14]equipment ↔ [15]appliances
[16]Boatswain ↔ [17]Bosun

There are others.

(a) navigating (b) stowage (c) maintenance
(d) responsibility (e) watch (f) sound

The Chief Officer is responsible for cargo and maintenance.
The Second Officer is responsible for navigation.
The Third Officer is responsible for the life-saving equipment.
The Boatswain is responsible for supervising the crew's work.
The Carpenter is responsible for sounding tanks and bilges.

## Applied terminology

### *(A) Terms relating to position in a ship*

*Exercise* 1.

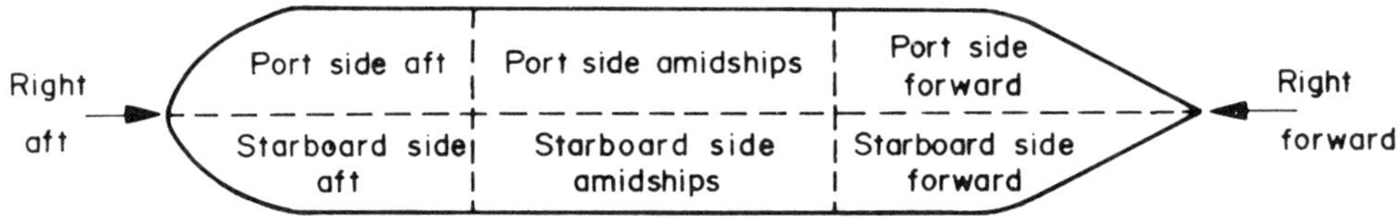


*Exercise* 2.

A modern general cargo ship has her engine room and bridge superstructure aft. She may have four holds forward of the bridge and one hold aft of the bridge. Forward of No. 1 hold is the forecastle and right forward is the jackstaff. Derricks are supported by masts and samson posts. They are stowed fore and aft when the ship is at sea. There are two lifeboats, one on the port side aft, another on the starboard side aft, abaft the funnel. The poop and the bridge superstructure are combined. There is an ensign staff right aft.

### *(B) Measurement*

*Exercise* 1.

(a) 0900 (b) 1330 (c) 1825 (d) 1745 (e) 1200
(f) 3.10 a.m. (g) 3.30 p.m. (h) 11.15 a.m.
(i) 8.45 p.m. (j) 11.55 p.m.

*Exercise* 2.

(a) 2000–midnight

(b) 0400–0800

(c) 1600–1800

(d) Middle Watch

(e) Forenoon Watch

(f) Second Dog Watch

(g) Chief Officer

(h) Second Officer

(i) Fourth Engineer

(j) Third Engineer

(k) 0600 (6.00 a.m.)

(l) 0830 (8.30 a.m.)

(m) 2300 (11.00 p.m.)

(n) 1930 (7.30 p.m.)

## Guided writing

### *(A) Description of the engine department*

The Engine Department is made up of a number of officers, petty officers and ratings. The Chief Engineer is in charge of the department and he is assisted by a Second Engineer, a Third Engineer and usually a Fourth Engineer, and sometimes an Electrical Officer. There are two petty officers: a Storekeeper and a Donkeyman. On tankers there is also a Pumpman. There are also a number of engine-room ratings including Firemen and Greasers. Sometimes there are Engineer Cadets too.

The Chief Engineer is responsible to the Master for the Engine Department. He also looks after the day-to-day running of the department. The Second Engineer is responsible for the maintenance of the engine room, deck and other machinery. Engine-room watchkeeping duties are the responsibility of the Second, Third and Fourth Engineers. The Second Engineer keeps the Morning and Evening watches; the Third Engineer keeps the Middle and Afternoon watches; the Fourth Engineer keeps the Forenoon and First watches. The repair and maintenance of all electrical equipment is the responsibility of the Electrical Officer. The Storekeeper and the Donkeyman are responsible to the Chief Engineer. The Storekeeper is responsible for the storeroom. The Donkeyman is responsible for lubrication. On tankers, the Pumpman is responsible to the Chief Officer for loading and unloading oil and water ballast. Of the engine-room ratings, Greasers are responsible for general oiling and cleaning duties, and Firemen are responsible for looking after the boilers.

# UNIT V SEAMANSHIP

## Reading comprehension

### (A) Ship handling

**Title:** Effects to be considered when handling a ship at sea or in narrow waters.

1. Propeller (a) axial thrust
   (b) transverse thrust } **under control**
2. Rudder
3. Wind } **not under control**
4. Tide

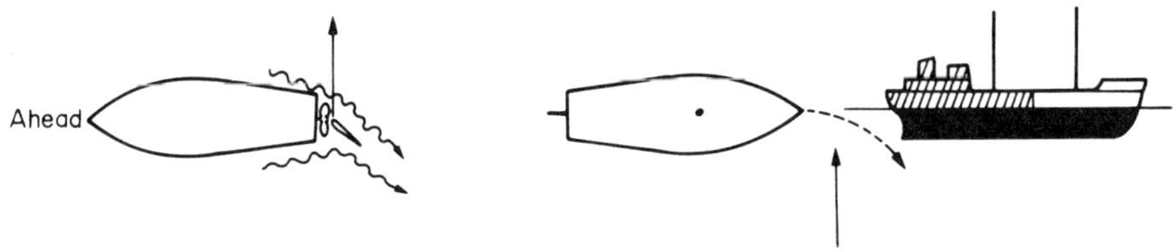


Rudder to port — More windage aft

## Grammar

### (B) Time relaters

*Exercise* 1.

(a) When a ship is completed, she undergoes speed trials.
(b) During construction, a ship is continually being examined.
(c) He did not go to sea until he was eighteen years old.
(d) On arriving on board, the passengers are shown to their cabins.
(e) During training, a cadet is paid by his company.
(f) The passengers did not go below until the sun went in.
(g) On being inspected, the hull was found to be damaged.
On inspection, the hull was found to be damaged.
(h) During speed trials, a vessel's maximum speed is measured.
(i) The pilot did not disembark until the ship was clear of the harbour.

## Reading comprehension

### (B) Ship handling (cont.)

**Leaving berth starboard side to (no wind or tide).** Single up to a backspring forward and a breastline aft. Put the engines to slow ahead and the rudder hard to starboard (1). Slack away on the breastline until the stern is clear of the berth (2). Stop engines, then go slow astern, put the rudder amidships. Let go the backspring and check on the breastline. The action of the breastline is to prevent transverse thrust taking the stern to port and consequently forcing the bows onto the quayside. When all is clear, stop engines and let go the breastline (3). Then put engines full ahead.

**Leaving berth tide astern.** Single up to a backspring forward and a breastline aft (1). Slack away on the breastline and the tide will force the vessel away from the berth. This can be helped by putting the rudder hard over away from the berth, in this case, hard to port (2). When clear of the berth, let go the backspring forward and the breastline aft, and put the engines half astern and the rudder amidships.

## Applied terminology

### *(A) Terms relating to a ship's movement*

*Exercise* 1.

(a) pitching (b) rolling (c) heaving
(d) surging (e) swaying (f) yawing

*Exercise* 2.

(a) A ship is said to be afloat when she is borne by the water.
(b) A ship is said to be underway when she is neither anchored, moored, made fast, nor aground.
(c) A ship is said to be making headway when she is moving forwards through the water.
(d) A ship is said to be making sternway when she is moving backwards through the water.
(e) A ship is said to be making leeway when she is moving sideways through the water being blown by the wind.
(f) A ship is said to overhaul another vessel when she is overtaking her.
(g) A ship is said to fall astern when she is dropping behind a faster vessel.
(h) A ship is said to be hove to when she is stopped at sea.
(i) A ship is said to be adrift when she is moving without means of propulsion.
(j) A ship is said to be aground when she is lying on the bottom or stuck on rocks.
(k) A ship is said to bear away when she is turned away from the wind.
(l) A ship is said to be moving broadside on when she is moving sideways through the water.

### *(B) Measurement*

*Exercise* 1.

(a) By the mark five
(b) And a half eight
(c) By the deep sixteen
(d) And a quarter seven

(e) A quarter less fourteen
(f) By the mark two
(g) And a quarter nine
(h) A quarter less four.

## Guided writing

### *(B) Description of weighing anchor*

When preparing to weigh anchor, it is necessary to take to the forecastle a hammer, an oil can and a hose. Before going to the forecastle, ask the engine room for power and water on deck. First, couple up the hose, lead it to the hawse pipe and open the water cock. Then, make sure that the windlass is out of gear and the brakes are on. Next, turn the windlass over slowly. While you are doing this, oil all the moving parts. After seeing that the gears are free to engage, put the windlass into gear. Then, remove the compressor bar. When the order to weigh anchor is received, release the brake and begin heaving in the cable. As the cable is heaved aboard, it is hosed down to wash away any dirt. When the shackles appear, the right number of rings is given on the forecastle bell. Ring the forecastle bell rapidly and haul down the anchor ball or anchor lights, when the anchor is aweigh. After the anchor is hove home, apply the brake and replace the compressor bar. Then take the windlass out of gear and inform the engine room that power and water are finished with. Finally, return all the gear.

(Other connecting words can also be used.)

# UNIT VI CARGO WORK

## Reading comprehension

### *(A) Different types of cargo*

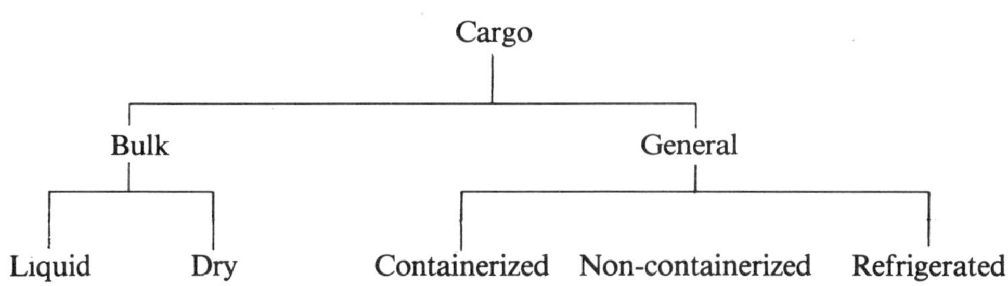

| Type of cargo | Examples | Type of ship | How (un)loaded |
|---|---|---|---|
| liquid bulk | crude oil, fuel oils | tankers | pumped through pipes |
| dry bulk | grain, iron-ore, coal, sugar | bulk carriers | conveyor belt, tubes, grabs |
| containerized | many types of general cargo | container ships | special cranes |
| non-containerized | tobacco, rubber, tea, rice | multideck vessels | cranes, derricks |
| refrigerated | meat, fruit, dairy produce | refrigerated ships | cranes, derricks |

## Grammar

### *(A) Prepositions of place*

*Exercise* 1.

(a) in (b) on (c) off (d) away from
(e) at (f) out of (g) in (h) on

*Exercise* 2.

The cargo for Liverpool and Glasgow is stowed in No. 2 hold. The cargo for Liverpool is stowed above the cargo for Glasgow, because it is to be unloaded first. Six drums of nitric acid to be unloaded at Liverpool are stowed on deck. Mail bags are stowed in the hatch square of the 'tween deck. In the lower hold cased goods are stowed on top of drums of caustic soda. Directly over the cased goods is earthenware.

### *(B) Relative clauses*

*Exercise* 1.

(a) Ships which carry coal are called colliers.
(b) Hemp and sisal rope are made from the leaves of a plant which grows in parts of Europe, Russia and the USA.
(c) The officer who is responsible for the Catering Department is called the Chief Steward.
(d) Cargoes which are classed as valuable are stowed in lock-ups.
(e) An officer who has the right qualifications and experience may become the Master of his own ship.

*Exercise* 2.

The Suez Canal, which took ten years to build, was opened on 17 November, 1869. It was designed by Ferdinand de Lesseps, who was a French engineer. The Canal, which is 105 miles long, joins the Mediterranean Sea to the Red Sea. Although it is twice as long as the Panama Canal, it cost half as much to build. This was because Suez is a sea-level canal from end to end, therefore locks, which are expensive to build, are unnecessary. The Canal, which was originally 58 metres wide at the surface and about 6 metres deep, has been widened and deepened several times since it was opened. It is now about twice its original breadth and depth. Until 1956 the Canal was operated by the Suez Canal Company. Since this date it has been operated by the Egyptian government, who plan to increase its depth to take supertankers.

## Reading comprehension

### *(B) Different types of cargo*

S2* Df; S3 N-Df; S4 N-Df; S6 N-Df; S7 Df; S9 N-Df; S10 Df; S12 N-Df; S14 N-Df; S17 Df; S19 Df; S21 Df; S22 Df; S23 Df; S26 Df; S28 N-Df; S30 Df, Df; S33 N-Df; S34 N-Df; S36 Df.

| | |
|---|---|
| 1 Merchant ships | ← 2 they |
| 2 bulk cargo | ← 3 The former |
| 2 general cargo | ← 4 The latter |
| 6 each item | ← 6 its |
| 8 tankers | ← 9 Most |
| 9 crude oil | ← 9 its |
| 10 The oil | ← 11 The cargo |
| 12 the ship | ← 12 her |
| 13 Dry bulk cargo | ← 14 The cargo |
| 14 The cargo | ← 15 It |
| 16 the cargo | ← 16 itself |
| 13 Dry bulk cargo | ← 16 the cargo |
| 18 non-containerized | ← 19 not in containers |
| 19 each commodity | ← 19 its |
| 23 Cargoes which are dusty | ← 23 them |
| 24 cargoes | ← 24 ones |
| 30 refrigerated holds | ← 31 These |

(a) Bulk cargo consists of a single cargo, which is usually carried loose.
(b) General cargo consists of a variety of goods, which are packed separately.
(c) Bulk cargo is carried in specially designed vessels.
(d) They may be of the reciprocating or the centrifugal type.
(e) Cargoes which have a strong odour will taint delicate cargoes.
(f) General cargoes are now being put into containers to help with the problem of stowage.
(g) Containers are 8 feet high and 8 feet wide ($2.44 \times 2.44$ m) and 20 feet or 40 feet (6.1 m or 12.2 m) in length.
(h) Perishable cargo is carried in refrigerated holds.

*S = sentence.

# Applied terminology

## (A) Terms relating to cargo handling gear

*Exercise* 1.

(a) A rope sling is used for lifting bags and bales.
(b) A canvas sling is used for lifting bags of grain, rice and coffee.
(c) A board sling is used for lifting bags of cement.
(d) A snotter is used for lifting cases, bales, hides and timber.
(e) A chain sling is used for lifting logs and iron rails.
(f) Plate clamps are used for lifting steel plates.
(g) Can hooks are used for lifting drums and barrels.
(h) A tray is used for lifting cases and drums.
(i) A box is used for lifting explosives.
(j) A net is used for lifting small packages and mail.
(k) A car sling is used for lifting cars and lorries.
(l) A heavy lifting beam is used for lifting locomotives and long heavy cargoes.

*Exercise* 2.

1st – Hatch beams
2nd – Beam bolts
3rd – Hatch covers
4th – Tarpaulins
5th – Hatch battens
6th – Wooden wedges.

## (B) Measurement

*Exercise* 1.

(a) Net register tonnage
(b) Underdeck tonnage
(c) Gross register tonnage

# Guided writing

## (A) Description of general cargo

General cargo, which is made up of an infinite variety of goods, can be classified according to its characteristics. For example, it can be described as wet, dry, clean or dirty; however, each cargo can have more than one of these characteristics. Because general cargo has different characteristics and forms of packaging, stowage can be a problem.

Wet cargoes, which include such cargoes as vegetable oil and wine, are carried in drums and barrels. There is always the possibility of leakage, therefore they must not be stowed on top of other cargoes.

Dry cargoes cannot possibly leak, but they can be damaged by leakage from wet cargoes, therefore they must be stowed away from them. Flour, animal feed and rice, which are carried in bags, are examples of dry cargoes.

Cargoes which leave a residue behind them are known as dirty cargoes. Because the dust from cement and fertilizers can damage other cargoes, holds must be cleaned out carefully after them.

Odorous cargoes, which include rubber and wool, give off fumes. These fumes can damage delicate cargoes, therefore they must be stowed separately.

Heavy cargoes must be given bottom stowage, because they will break more fragile cargoes. When very heavy items are carried in ships, they should be stowed over bulkheads, as they might damage the deck plating.

Cargoes which are combustible, toxic or liable to explode are classified as dangerous cargoes. Because they are a risk to human life and the safety of the ship, they are subject to strict laws.

Perishable food is usually carried in refrigerated ships; however, some general cargo vessels have refrigerated holds. Meat and fish are often frozen solid, but fruit and vegetables are only chilled.

(*Note:* There are other ways of joining some of these sentences.)

# UNIT VII NAVIGATION

## Reading comprehension

### *(A) Navigating techniques and instruments*

1. Different techniques of navigation
   (a) celestial navigation
   (b) coastal navigation (pilotage)
   (c) dead reckoning

Instruments used in navigation
(a) a sextant
(b) a chronometer
(c) a compass

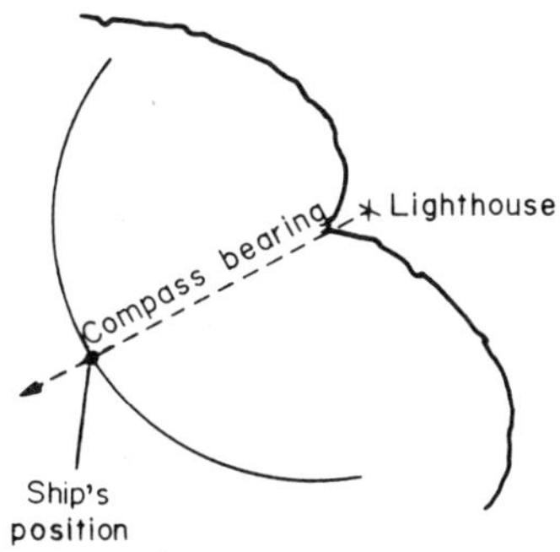


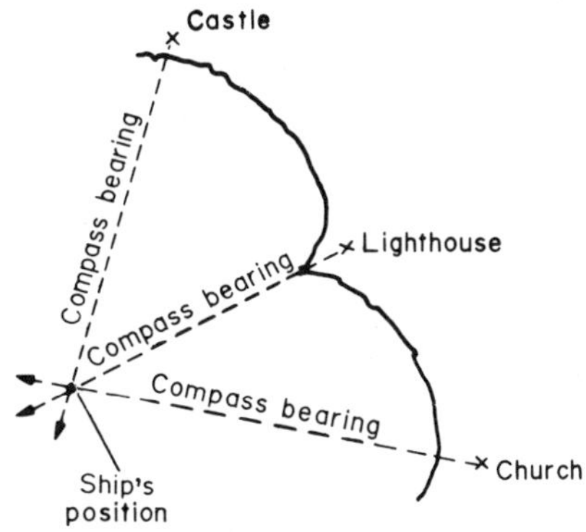


# Grammar

## (A) Prepositions of places

*Exercise* 1.

On 10 September, 1980 the bulk oil carrier, the SS *Enterprise*, left the port of Nagasaki in Japan for Kharg Island in the Gulf. She steamed down the east coast of China, through the Taiwan Straits and across the south China Sea calling at Singapore. She then sailed through the Straits of Malacca and across the Indian Ocean to Kharg Island.

On 16 June, 1980 the passenger liner, the MV *Orient*, left the port of Southampton in the UK for Naples in Italy. She steamed down the English Channel, across the Bay of Biscay and down the west coast of Portugal calling at Lisbon. She then sailed through the Straits of Gibraltar and across the Mediterranean Sea to Naples.

## (B) Relative clauses

*Exercise* 1.

(a) Ships using the Panama Canal pay canal dues.
(b) Rope made of nylon does not rot.
(c) An Officer with/having a Master's Certificate can hope to become the master of his own ship one day.
(d) Cannot be reduced.
(e) A sailor working aloft should wear a safety harness.
(f) Steel containing more than 0.7 per cent carbon has a high tensile strength.
(g) Cargoes with/having a low stowage factor are usually put at the bottom of the hold.
(h) Cannot be reduced.
(i) Ships classed by Lloyds have to undergo strict tests.
(j) The sailors sitting at the back of the coach were hurt in the accident.

*Exercise* 2.

(a) Copper and tin, from which bronze is made, are non-ferrous metals.

(b) Pallets, on which crates and cases are stacked, are pieces of cargo-handling equipment.

(c) The windlass, by which the anchors are raised and lowered, is found on the forecastle.

(d) Ventilators, through which air flows, are positioned over the cargo holds.

(e) Hatchways are covered by hatch covers, over which tarpaulins are then spread.

(f) A sextant is a navigating instrument with which a Deck Officer takes observations of the sun.

(g) Speed trials, during which the ship's maximum speed is measured, take place when the ship is out at sea.

(h) The National Union of Seamen, to which all Merchant Mavy ratings must belong, is an organization which looks after the interests of seamen.

(i) The flight deck is an open deck on an aircraft carrier from which planes and helicopters take off.

## Reading comprehension

### *(B) Electronic aids to navigation*

Electronic devices used for navigation:

(a) the echo sounder
(b) the sonar system
(c) radar
(d) the radio direction finder
(e) Decca, Omega, Loran
(f) satellites
(g) sensors

## Applied terminology

### *(A) Terms relating to compass cards*

*Exercise* 1.

(a) North
(b) North-East
(c) East
(d) South-East
(e) South
(f) South-West
(g) West
(h) North-West

*Exercise* 2.

(a) S – 180°
(b) NE – 045°

(c) NW – 315°
(d) E – 090°
(e) SW – 225°
(f) N – 0°
(g) W – 270°
(h) SE – 135°

*Exercise* 3.

(a) Ship E is abaft the starboard beam.
(b) Ship F is on the starboard quarter.
(c) Ship G is (dead) astern/Ship G is astern of the tanker.
(d) Ship H is on the port quarter.
(e) Ship I is abaft the port beam.
(f) Ship J is abeam./Ship J is on the port beam.
(g) Ship K is before the port beam.
(h) Ship L is on the port bow.

### *(B) Measurement*

*Exercise* 1.

(a) 936
(b) 2537
(c) 2135
(d) 1319
(e) 1950
(f) 2719

*Exercise* 2.

| MV *Panama* | Distance | Time | Speed |
|---|---|---|---|
| Monday | 180 N miles | 12 hours | 15 knots |
| Tuesday | 336 N miles | 24 hours | 14 knots |
| Wednesday | 288 N miles | 24 hours | 12 knots |
| Thursday | 52 N miles | 8 hours | 6.5 knots |
| Friday | 110 N miles | 10 hours | 11 knots |
| Saturday | 312 N miles | 24 hours | 13 knots |
| Sunday | 152 N miles | 16 hours | 9.5 knots |

## Guided writing

### *(A) Description of a voyage*

On 10 May the oil tanker the MV *Voyager*, which was carrying a cargo of petroleum products, left Grangemouth in Scotland for the ports of Gothenburg, Stockholm and Luleå in Sweden. Grangemouth is a small port on the River Forth, near which are important oil refineries. She sailed down the Firth of Forth and passed under the Forth Road Bridge and the Forth Rail Bridge and out into the North Sea. Her passage across the North Sea was good because the weather was clear and the wind was light. She sailed through the Skagerrak and arrived at Gothenburg, which is a port on the west coast of Sweden, on 11 May. At Gothenburg she discharged gas oil and loaded petrol. She left Gothenburg the next day and sailed down the Kattegat and between the islands of Fyn and Zealand and through the Fehmarn Belt into the Baltic Sea. She then sailed north for Stockholm passing between Bornholm and the mainland and going west of Gotland. On 15 May she arrived at Stockholm, which is the capital of Sweden. After unloading part of her cargo at Stockholm she sailed for Luleå the next day. She sailed north up the Gulf of Bothnia and arrived at Luleå on 19 May. Luleå, from which iron-ore is exported, is only open during the summer months. After unloading the rest of her cargo, she set sail for Grangemouth, but on her way through the Kattegat, she was ordered to proceed to Hamburg, which is a port in Germany.

(*Note:* there are other ways of describing the voyage.)

# UNIT VIII MAIN ENGINES

## Reading comprehension

### *(A) Different types of marine engines*

Marine engines

- Diesel engine
  - Slow-speed
  - Medium-speed
- Steam turbine
- Gas turbine
- Nuclear plant

| Type of engine | Application |
|---|---|
| slow-speed diesel | tankers, bulk carriers |
| medium-speed diesel | cargo liners, tankers, bulk carriers |
| steam turbine | tankers, container ships, cargo oil pumps |
| gas turbine | naval vessels, container ships |
| marine nuclear plant | naval vessels, submarines, some cargo ships |

## Grammar

### *(A) Causal verbs*

*Exercise* 1.

(a) The Second Mate got the cadet to plot the course.
(b) The First Mate had the cargo checked.

(c) The Bosun had the boats lowered by the apprentices.
(d) The Carpenter got the bilges cleaned by the ratings.
(e) The Master got the compasses adjusted.
(f) The Chief Engineer had the Second Engineer supervise the work.
(g) The Second Engineer got the greasers to lubricate the machinery.
(h) The Chief Steward had the saloon swept by the galley boy.

*Exercise* 2.

(a) The carpenter shortened the plank.
(b) The gap between the two vessels widened.
(c) You need to sharpen your pencil before doing chartwork.
(d) Constant pressure weakens a spring.
(e) Speed was reduced to lighten the load on the engine.
(f) The ropes had to be lengthened because they would not reach.
(g) He had great difficulty in loosening the nut.
(h) Adding carbon to steel hardens it.
(i) Reducing the carbon content softens it.
(j) The hole had to be enlarged before the bolt would fit.
(k) Friction is increased by roughening a surface.

## *(B) Cause and effect*

*Exercise* 1.

(a) Poor lubrication produces friction.
(b) Instability can be caused by bad loading.
(c) Overloading leads to structural strain.
(d) Efficiency results from good management.
(e) Heat loss leads to a reduction in power.
(f) Carelessness often causes accidents.
(g) Careful navigation leads to fuel economy.
(h) Lack of maintenance causes breakdowns.

(*Note:* other answers are also possible.)

*Exercise* 2.

(a) The piston is driven downwards, causing the flywheel to rotate.
(b) The flywheel rotates, making the pin travel downwards.
(c) The pin travels downwards, pulling the string.
(d) The string is attached to the drum in such a way that it causes the drum to rotate.
(e) When the piston travels upwards, the tension on the string is reduced, allowing the drum to return to its original position by means of a spring.
(f) Therefore each time the flywheel rotates, it causes the drum to rotate backwards and forwards.

(g) The point of a pencil is held against the paper on the drum, consequently the pencil marks the length of the stroke on the paper each time the drum rotates.
(h) The upward movement of the piston causes compression of the gas in the cylinder.
(i) The pressure of the gas raises the indicator piston.
(j) The indicator piston rises, compressing the spring.
(k) The piston is attached to the arm holding the pencil, consequently the pencil moves upwards, indicating the rise on the paper.

## Reading comprehension

### *(B) Two-stroke and four-stroke cycles*

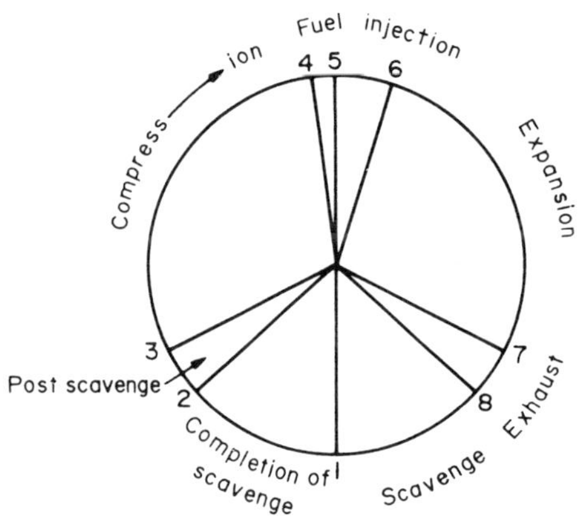


Two-stroke cycle

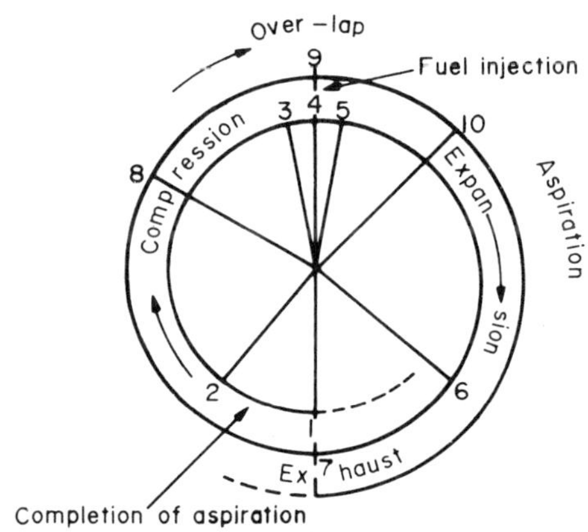


Four-stroke cycle

## Applied terminology

### *(B) Measurement*

*Exercise* 1.

| Quantity | Unit | Symbol |
|---|---|---|
| area | square metre | $m^2$ |
| volume | cubic metre | $m^3$ |
| velocity | metre per second | m/s |
| angular velocity | radian per second | r/s |
| acceleration | metre per second second | $m/s^2$ |
| density | kilogramme per cubic metre | $kg/m^3$ |
| momentum | kilogramme metre per second | kg m/s |

*Exercise* 2.

| Quantity | Unit | Symbol |
|---|---|---|
| torque | newton metre | Nm |
| stress | newton per square metre | $N/m^2$ |
| pressure | newton per square metre | $N/m^2$ |
| intensity of heat flow rate | watt per square metre | $W/m^2$ |
| thermal conductivity | watt per metre degree celsius | W/m °C |
| coefficient of heat transfer | watt per square metre degree celsius | $W/m^2$°C |
| heat capacity | joule per degree celsius | J/°C |
| specific heat capacity | joule per kilogramme degree celsius | J/kg °C |

## Guided writing

### (A) *Description of an engine governor*

The diagram above shows a hydraulic engine governor which may be fitted to control the speed of an auxiliary diesel engine. The governor incorporates two systems: the one consisting of the mechanical ballhead, which senses any change in the engine speed; the other consisting of the hydraulic piston valve. It also consists of the power piston, which operates the fuel pump control setting. Any alteration in speed setting can be made by altering the speed adjustment control.

When the engine speed falls below the desired limit, the driveshaft rotates slower, making the flyweights move in. The flyweights move in, lowering the piston valve. This causes the oil pressure to increase, raising the power piston and compressing the return spring. The fuel control valve opens, increasing the fuel and consequently the engine speeds up.

When the engine speed rises above the desired limit, the driveshaft rotates faster, making the flyweights move out. The flyweights move out, raising the piston valve. This causes the oil pressure to decrease, allowing the return spring to extend, forcing the power piston down. The fuel control valve closes, decreasing the fuel and consequently the engine slows down.

(*Note:* There are other ways of joining these statements.)

# UNIT IX AUXILIARY MACHINERY

## Reading comprehension

### (A) *Functions of auxiliary machinery*

Auxiliary machinery is designed:

to supply the needs of the main engines and boilers – air compressors, coolers, feed water heaters to keep the ship dry and trimmed – bilge pumps, ballast pumps

to supply domestic needs – distillation plant, sewage plant, heaters, air-conditioners

to apply the main power of the engines for propulsion and manoeuvring – thrust shaft, intermediate shafts, propeller shaft, propeller, rudder, steering gear

to supply the ship with electrical power and lighting – generators

to moor the ship and handle cargo – windlass, capstans, winches, cranes, cargo oil pumps

to provide for safety – fire fighting equipment, fire detection equipment, lifeboat engines, launching gear

## Grammar

### *(A) Change of state verbs*

*Exercise* 1.

(a) converts
(b) get/become
(c) liquefies
(d) is converted into
(e) becomes
(f) rot
(g) turns
(h) become
(i) turns

### *(B) Comparisons*

*Exercise* 1.

(a) A statute mile is shorter than a nautical mile.
(b) Sisal rope is cheaper than manila rope.
(c) A petrol engine is less economical than a diesel engine.
(d) 100°F is cooler than 100°C.
(e) The Second Officer is junior to the Chief Officer.
(f) There are fewer members of the Catering Department in a cargo vessel than in a passenger liner.
(g) Water freezes at a higher temperature than mercury.
(h) Mild steel contains less carbon than high carbon steel.
(i) A VLCC is smaller than a ULCC.
(j) A conventional submarine is slower than a nuclear submarine.

*Exercise* 2.

(a) A statute mile is not as/so long as a nautical mile.
(b) Sisal rope is not as/so expensive as manila rope.
(c) A petrol engine is not as/so economical as a diesel engine.
(d) 100°F is not as/so hot as 100°C.
(e) Low carbon steel is not as/so hard as high carbon steel.
(f) Not as/so many tankers will be built this year as last year.
(g) Water does not have as/so low a freezing point as mercury.
(h) A single deck vessel does not have as/so many decks as a 'tween deck vessel.

(i) Before British ships did not carry as/so much tonnage as nowadays.
(j) A conventional submarine does not have as/so high an underwater speed as a nuclear submarine.

## Applied terminology

### (A) Terms relating to ancillary services

*Exercise* 1.
drain tank, strainer, Pumps, filters, cooler, distribution branches, filters, vent, level gauge, pump, drain tank, heater, centrifuge, drain tank, strainer.

### (B) Measurement

*Exercise* 1.
(a) Volts equal amperes multiplied by ohms.
(b) Coulombs equal amperes multiplied by seconds.
(c) Volts equal watts divided by amperes.
(d) Ohms equal volts divided by amperes.
(e) Farads equal amperes multiplied by seconds divided by volts.
(f) Henrys equal volts multiplied by seconds divided by amperes.

*Exercise* 2.

| | | | |
|---|---|---|---|
| 1,000,000 J | = one megajoule | = 1 MJ | = $10^6$ J |
| 1,000 N | = one kilonewton | = 1 KN | = $10^3$ N |
| 0.001 V | = one millivolt | = 1 mV | = $10^{-3}$ V |
| 0.000001 Ω | = one microhm | = 1 μΩ | = $10^{-6}$Ω |
| 1,000 J | = one kilojoule | = 1 KJ | = $10^3$ J |
| 0.001 A | = one milliampere | = 1 mA | = $10^{-3}$ A |
| 1,000,000 W | = one megawatt | = 1 MW | = $10^6$ W |
| 1,000 V | = one kilovolt | = 1 KV | = $10^3$ V |
| 0.01 m | = one centimetre | = 1 cm | = $10^{-2}$ m |
| 0.000001 m | = one micrometre | = 1 μm | = $10^{-6}$ m |

## Guided writing

### (A) Description of types of pump

The pumps which are used on board ship can be divided into two main groups: displacement pumps and centrifugal pumps.

In displacement pumps the volume of the pump chamber is increased by raising the piston. This causes a vacuum into which the liquid is drawn from the suction pipe. The piston is then lowered, decreasing the volume of the pump chamber and forcing the liquid out into the delivery pipe.

Centrifugal pumps consist of an impeller, which rotates at high speed, inside a casing. The liquid, which enters through a suction pipe at the centre, is thrown by centrifugal force against the surrounding casing. It is then discharged through the delivery outlet.

A simple kind of reciprocating displacement pump is the single-acting ram pump, which consists of a ram moving up and down inside a chamber fitted with a non-return suction valve and a non-return delivery valve. When the piston moves up, a vacuum is formed in the chamber into which liquid is drawn through the suction valve. Then the piston moves down, creating a force on the liquid which closes the suction valve and forces the liquid out through the delivery valve.

Another type of reciprocating pump is the double-acting piston pump, which works on the same principle as the single-acting pump. The chamber is fitted with suction and return valves at the top and bottom, consequently the liquid can be drawn in and discharged on each stroke.

The gear-wheel pump, which is used for pumping lubricating and fuel oils, is an example of a rotary displacement pump. It consists of interlocking gear wheels which rotate, each tooth on both wheels leaving a vacuum for liquid to flow into. The next tooth to enter forces the liquid out. Therefore, the liquid enters a space, is carried round and then forced out into the delivery tube.

# UNIT X MAINTENANCE

## Reading comprehension

### *(A) Maintenance*

(a) Metal primers, undercoats, top coats, heat-resistant paints, non-slip paints, varnishes, topside paint, boot-topping paint, anti-corrosive paint, anti-fouling paint.
(b) The Chief Officer
(c) 1st – Clean and wash area with a solution to remove salt, dirt, and oil.
2nd – Rinse with fresh water to remove all trace of the cleaning solution.
3rd – Remove loose paint, scale and rust with a chipping hammer.
4th – Brush area over with a wire brush.
(d) Cement wash, bitumen or paint.

# Grammar

## (A) Noun compounds

*Exercise* 1.

(a) a valve shaped like a needle
(b) steel which contains carbon
(c) a tank for petrol
(d) a rope made of wire
(e) bolts at the corners
(f) a pump operated by foot
(g) plates on/in the floor
(h) a hatch invented by MacGregor
(i) an engine which runs on diesel
(j) a ship for carrying containers

*Exercise* 2.

(a) compressed air
(b) lubricating oil
(c) mooring ropes
(d) a riveted joint
(e) a machined surface
(f) a sticking valve
(g) heated oil
(h) scavenging air
(i) welded joint
(j) cleaning rags

## (B) Purpose links

*Exercise* 1.

Suggested answers:

(a) Ships carry lifeboats so that the crew can escape if the ship sinks.
(b) Engines must be regularly serviced in order to prevent breakdowns.
(c) He opened the window to let in some fresh air.
(d) Safety valves are fitted so as to prevent accidents and serious damage.
(e) Ships are taken on sea trials so that they may demonstrate their capabilities and reveal any faults.
(f) Goggles should be worn so as to protect the eyes.
(g) He set his alarm clock so that he would not oversleep.
(h) Warships are painted grey in order to camouflage them.

# Reading comprehension

## (B) Fault chart

(a) The engine would refuse to start.

(b) Put on heating steam and circulate fuel lines.
(c) At the starting air pressure gauge.
(d) When the injection timing is wrong.
(e) Turn on No. 2 filter and clean No. 1 filter.
(f) The engine may refuse to pick up firing after starting.
(g) Refill the service tank and prime the fuel pumps.
(h) At the engine governor, governor links and fuel pump.
(i) The engine would work irregularly.
(j) At all the gauges.

## Applied terminology

### *(A) Terms relating to maintenance*

*Exercise* 1.
(a) a hammer designed by Munday
(b) a scraper which is straight in shape
(c) a scraper which is shaped like a goose's neck
(d) a scraper with a long handle
(e) a brush made of wire
(f) a scraper with three corners used for scraping wood
(g) a broom for sweeping decks
(h) a brush shaped like a Turk's head

*Exercise* 2.
(a) a brush which is flat in shape
(b) a brush which is oval in shape
(c) a brush for tar
(d) a brush used for painting lines
(e) a brush used for stencils
(f) a brush shaped like and used like a pencil

### *(B) Measurement*

*Exercise* 1.
(a) a ruler which is six inches long
(b) a vessel with two propellers
(c) a nail which is two inches long
(d) a stand with two legs
(e) a hemp line made up of three strands
(f) an oil designed to mix with different grades of oil
(g) a derrick designed to lift up to 15 tons
(h) a scavenging air system designed to flow one way
(i) coils of rope 120 fathoms long

## Guided writing

There are a number of reasons for wear in cylinder liners. Wear may be due to friction, corrosion and abrasion. Each of these may have a number of causes.

Frictional wear depends on the materials used in the liner and in the piston rings. It takes place between the sliding surface of the cylinder liner and the piston rings. It may be caused by inefficient lubrication of the cylinder or by overloading the engine. In addition, engine operating conditions also affect frictional wear. For example, frictional wear may increase, if the air and fuel are contaminated.

Having discussed the causes of frictional wear, let us go on to consider the reasons for corrosion. Corrosion occurs when heavy fuels with a high sulphur content are burned. It is caused by acids which are formed during combustion. Sulphuric acid corrosion may be caused if the cooling water temperature is too low. This allows moisture to condense in the cylinder cooling jacket and consequently sulphuric acid may form in the cylinder.

Now let us look briefly at the causes of abrasion. Abrasion is caused by hard particles. These may be the product of combustion or mechanical wear. Hard particles may also be produced by corrosion.

We have shown, then, that cylinder liner wear is caused by friction between the liner and the piston rings, by corrosion – mainly from burning heavy fuels – and by abrasion from the products of wear, corrosion and combustion.

# UNIT XI SAFETY ABOARD

## Reading comprehension

### *(A) Collision Regulations 1977 (extracts)*

(a) The black vessel should keep out of the way of the vessel being overtaken.
(b) The black vessel should alter course to starboard.
(c) The black vessel should keep out of the way and avoid crossing ahead of the other vessel.
(d) The black vessel should maintain her course and speed.
(e) The black vessel must take action to avoid collision.
(f) The black vessel must keep out of the way of the other vessel as she is a sailing vessel.

# Grammar

## (A) Obligation

*Exercise* 1.

2. Report by telephone at regular intervals to the duty Deck Officer.
3. Do only the task you have been told to do.
4. Get help if necessary.
5. Inform the Engineer Officer in charge when you leave the space.
6. Keep the space well-lit at all times.
7. Keep the space clean.
8. Test all alarm circuits regularly.
9. Be on guard at all times against the sudden starting of automated machinery.

## (B) Conditionals

*Exercise* 1.

(a) If the weather is bad, we will work below deck.
(b) If we had a torch, we would be able to see.
(c) If I had passed the exam, I would have been promoted.
(d) If you touch that wire, you will be electrocuted.
(e) If I saw him again, I would recognise him.
(f) If the car had been cheaper, I would have bought it.

# Reading comprehension

## (B) Code of safe working practices (extracts)

*Exercise* 1.

| Part of the body | Type of protection | Situations |
|---|---|---|
| body | well-fitting clothes | working in machinery spaces |
| neck | no ties, no sweat rags | working near machinery in motion |
| fingers | no rings | working in machinery spaces |
| eyes | goggles | handling chemicals, grinding, drilling, scaling, working a lathe etc. |
| feet | industrial footwear | working in machinery spaces |
| head | helmet | overhauling machinery |
| ears | ear defenders | working where noise level is high |

*Exercise* 2.

If you are working in machinery spaces, you should not wear rings, in order to protect your fingers. (N.B. Rings can also fall off into moving parts.)

If you are handling chemicals, grinding, drilling, scaling, working a lathe, etc., you should wear goggles to protect your eyes.
If you are working in machinery spaces, you should wear industrial footwear to protect your feet.
If you are overhauling machinery, you should wear a helmet to protect your head.
If you are working where noise level is high you should wear ear defenders.

## Applied terminology

### (A) Terms relating to life-saving apparatus

(a) lifeboat
(b) lifebuoy
(c) life-jacket
(d) life-raft

### (B) Measurement

*Exercise* 1.

(a) $^1/_9$ tonnes
(b) $^1/_6$ tonnes
(c) $^6/_{25}$ tonnes
(d) $^3/_8$ tonnes
(e) $^8/_9$ tonnes
(f) $^8/_{45}$ tonnes
(g) $^{49}/_{50}$ tonnes
(h) $1\,^1/_2$ tonnes

## Guided writing

### (A) Description of lowering a lifeboat

First, the coxswain orders the boat's crew to fall in and number off. Then, he details members of the crew to do particular jobs. Next, the harbour pins, which prevent the boat being accidentally lowered, are cleared. After that, two crew members enter the boat and ship the plug, which is made of cork and fits in a hole near the fore-and-aft line of the boat. Next, the toggle painter, which prevents the boat from drifting astern, when lowered, is secured to the ship. Before the boat is lowered to the embarkation deck, the gripes, which secure the boat when in the stowed position, are cleared away. The bowsing tackles are secured and the boat bowsed in. The tricing pennants, which bring the boat towards the embarkation deck as it is lowered, are slipped because the bowsing tackles take the weight of the boat off the tricing pennants. The passengers and the crew now embark. Then, the bowsing tackles are released and the boat is lowered to the water.

(*Note:* other versions are possible.)

# UNIT XII COMMUNICATION AT SEA

## Grammar

### (A) Modal expressions of certainty

*Exercise* 1.

(a) may
(b) must
(c) might/could
(d) may not have
(e) will
(f) may
(g) must have
(h) will not

*Exercise* 2.

(a) is likely to
(b) is certain/sure to
(c) may possibly
(d) may possibly not have
(e) is certain/sure to
(f) are likely to
(g) is certain/sure to have
(h) are unlikely to

(*Note:* There are other possible answers in these exercises, depending how certain you are in each situation.)

### (B) Non-finite clauses

*Exercise* 1.

(a) Being the senior officer left alive, the Lieutenant took command of the vessel.
(b) Although damaged, the radio still functioned.
(c) Having painted the port side, the seamen started on the starboard side.
(d) When repaired, the ship will be put up for sale.
(e) Since joining the Royal Navy, he has visited many countries.
(f) The book to read on that subject is in the library.
(g) Painted, it would look as good as new.
(h) Being wet, the ropes were left on the deck to dry.
(i) They continued rowing, though exhausted.
(j) When opened, the container should be emptied immediately.

*Exercise* 2.

(b) He had no money, therefore he had to walk into town.
(c) The ship slowed down, because the fog had become thicker.
(d) If you take this medicine, you will feel better.
(e) They searched carefully, but there was no-one there.
(f) He washed the dishes, then dried them.
(g) If you work hard, you will pass your examinations.
(h) There is a choice of red wine or white wine.
(i) He ran on deck, because the alarm bells were ringing.
(j) He tried hard, but just failed.

# Applied terminology

## (B) Measurement

*Exercises* 1 *and* 2.

| Frequency band | Frequency range | Typical uses |
|---|---|---|
| very low frequencies (vlf) | 3–30 kHz | time signals, long distance |
| low frequencies (lf) | 30–300 kHz | mrt mob, nav aids, fxd |
| medium frequencies (mf) | 300–3000 kHz | mrt mob, bc |
| high frequencies (hf) | 3–30 MHz | fxd, mrt & aer mob, bc, amtr |
| very high frequencies (vhf) | 30–300 MHz | tv&r bc, rdr, mrt & aer mob, sh-wve, bc, fxd, mob |
| ultra high frequencies (uhf) | 300–3000 MHz | fxd, mob, mrt & aer mob, bc, nav, met, sp, shrt-dst |
| superhigh frequencies (shf) | 3–30 GHz | rdr, rad rel, nav, sp, sat comm, exp |

# Guided writing

## (A) Description of radio communication system

A basic radio communication system consists of a microphone or key, a transmitter and transmitting aerial, a receiver and receiving aerial, and a loudspeaker or headphones. The microphone or Morse key produces an electrical signal. The radio transmitter generates radio-frequency waves which vary in accordance with the electrical signal. The transmitting aerial radiates these waves into space and the receiving aerial receives them. The radio receiver amplifies the signal and reproduces the original electrical signal. The loudspeaker or headphones convert this signal into sound, consequently the original sound is reproduced.

In a radio communication system the information to be transmitted must first be converted into an electrical signal. In radio-telegraphy this is done using a Morse key, whereas in radio-telephony this is done using a microphone. Then, this information is superimposed on radio waves by the transmitter, using the signal from the Morse key or microphone. These radio waves are generated by the transmitter at a definite frequency. After that, they are conveyed to the transmitting aerial for radiation. The receiving aerial picks up the radio signal in addition to a mass of other signals as well. The radio signal is then dealt with by the radio receiver. The receiver must do a number of things. After selecting the desired signal, it must amplify it. After that, it must reproduce the original electrical signal, which is then sent to headphones or a loudspeaker. These reproduce the original sound and, therefore, complete the system.

# UNIT XIII RADIO COMMUNICATIONS

## Grammar

### *(A) Message types*

*Exercise* 1.

(a) Warning
(b) Question
(c) Information
(d) Instruction
(e) Question
(f) Advice
(g) Intention
(h) Request
(i) Instruction
(j) Warning

### *(B) Message patterns*

*Exercise* 1.

(a) **Information:** Dredging operations are completed in the fairway. Over.
**Information – received:** Dredging operations are completed in the fairway. Over.
(b) **Intention:** I intend to anchor now. Over.
**Intention – received:** You intend to anchor now. Over.
(c) **Question:** Is buoy number: two-three in the correct position? Over.
**Answer:** Negative, buoy number: two-three is not in the correct position. Over.
(d) **Request:** Please send welding equipment, reason: urgent repair.
**Information:** We have qualified welders on board. Over.
(e) **Advice:** Please proceed to Avonport, reason: there is cargo for you. Over.
**Advice – received:** Proceed to Avonport, reason: there is cargo for me. Over.
(f) **Warning:** A strong easterly wind is blowing at the berth. Over.
**Warning – received:** A strong easterly wind is blowing at the berth. Over.
(g) **Instruction:** Turn starboard immediately. Over.
**Instruction – received:** Turn starboard immediately, negative, reason: sailing vessel on my starboard bow. Over.

*Exercise* 2.

(a) **Instruction:** Keep clear of the approach channel, reason: large vessel leaving.
(b) **Intention:** I intend to attempt rescue by Breeches-buoy.

(c) **Question:** What is your ETA at the channel entrance?
(d) **Request:** Please confirm that your length is two-zero-zero metres?
(e) **Advice:** Please pass ahead of me, reason: I am slowing down.
(f) **Answer:** Negative, shore-based radar assistance is not available.
(g) **Warning – received:** I'll keep a look out for wreckage.

## Reading comprehension

### *(B) VHF procedures*

*Page 191*

| Station making the call | Station responding to the call |
|---|---|
| 1. 'Polar Star, Nine Victor Alpha Tango;<br>Polar Star, Nine Victor Alpha Tango;<br>This is Rambler, Golf Xray Xray Xray;<br>Rambler, Golf Xray Xray Xray<br>On VHF Channel one-six.<br>Over' | |
| | 2. 'Rambler Golf Xray Xray Xray<br>This is Polar Star, Nine Victor Alpha Tango.<br>Over' |
| 3. 'Polar Star This is Rambler.<br>Switch to VHF channel zero-six.<br>Over.' | |
| | 4. 'Rambler. This is Polar Star<br>Agree VHF channel zero-six<br>Over' |
| 5. 'Polar Star. This is Rambler<br>Question: What is your destination?<br>Stay on.<br>Over' | |

6. 'Rambler. This is Polar Star

Answer: My destination is Dakha.

Over'

7. 'Polar Star. This is Rambler

Understood, Dakha.

INFORMATION : Dakha pilot services

suspended, reason: gales.

Over'

8. 'Rambler. This is Polar Star.

INFORMATION RECEIVED: Dakha pilot

services suspended, reason: gales

Over.'

9. 'Polar Star. This is Rambler

Nothing more.

Out'

## Applied terminology

### *(A) Terms relating to VHF communication*

*Exercise* 1.

(a) Pan-Pan (b) Pan-Pan (c) Sécurité
(d) Mayday (e) Sécurité (f) Pan-Pan

*Exercise* 2.

(a) 'Mayday Mayday Mayday
This is
Rattler Rattler Rattler
Mayday
Rattler, Golf Xray Xray Xray

(b) 'Pan-Pan Pan-Pan Pan-Pan
This is
Vega Vega Vega
Pan-Pan
Vega, Seven Victor Alfa Tango

Position: latitude: five-zero degrees three-zero minutes North, longitude: zero-three-nine degrees two-zero minutes West.
Collision with iceberg, sinking.
Request immediate assistance.
Over.'

Position: bearing zero-nine-zero degrees true, distance: one-five miles from Ras Sarkan.
Lost propeller.
Require tow.
Over.'

*Exercise* 3.

'Sécurité Sécurité Sécurité
All ships All ships
This is
Nitonradio Nitonradio Nitonradio
Sécurité
Nitonradio
Navigational warning; military exercise in sea area **Delta**.
Switch to VHF channel zero-six.
Over.'

'Sécurité Sécurité Sécurité
All ships All ships
This is
Nitonradio Nitonradio Nitonradio
Sécurité
Nitonradio
Navigational warning; military exercise involving darkened ships, submarines and aircraft taking place in sea area **Delta** from 0800 hours GMT, March 3rd. to 2200 hours GMT, March 7th.
I say again.
Military exercise involving darkened ships, submarines and aircraft taking place in sea area **Delta** from 0800 hours GMT, March 3rd. to 2200 hours GMT, March 7th.
Out.'

## *(B) Measurement*

*Exercise* 2.

(a) **Request:** Please deliver, quantity: five-zero-zero metric tonnes, fresh water to berth number: one-eight. Over.

(b) **Request:** Please supply welding equipment and quantity: two sets of gas bottles. Over.

(c) **Request:** Please deliver, quantity: one-zero litres clear varnish, quantity: five-zero litres deck paint, green. Over.

(d) **Request:** Please deliver, quantity: one replacement scanner motor, part number: one-one-zero-Victor-Delta-Charlie-two-five-zero-four-zulu, for Decca Radar Transar, number: eight-zero-nine. Over.

## Guided writing

### (A) Keeping the radiotelephone log

*Page 199*

Name of Vessel: Essex Coast. Callsign: GMML
Bound for Leith
Radio Officer on watch: J. Smith
Watch began: 08.30
Silence period observed: 08.30-08.33

Suggested paraphrases:

(a) I have 1 radiotelegram and a TR for you. I will send these on 2016KHz. Over.
(b) Received. I agree 2016KHz. Listen for me on 1856KHz. Over.
Received. Change frequency now.
(c) How do you read me? Over.
(d) The intelligibility of your signal is 5. I am ready. Over.
(e) Transit Record Essex Coast GMML leaving Aberdeen bound for Leith. Over.
(f) Received TR. Over.

## Note-taking practice

THE MARCONI INTERNATIONAL MARINE CO. LTD.
A GEC—Marconi Electronics Company
MARCONI MARINE
**MARCONIGRAM**
(BY RADIO/TELEGRAPH/TELEX SERVICE)
THE QUEEN'S AWARD FOR EXPORT ACHIEVEMENT

| Prefix | Office of Origin | No. | No. of Words | Date filed | Time filed |
|---|---|---|---|---|---|
| MSG | ESSEX COAST | 1 | 16 | 12 | 0830 |
| Service Instructions | | Sent by | Sent to | Date Sent | Time Sent |
| GBØ8 | | | | | |

| CHARGES | | |
|---|---|---|
| Mobile Station | | |
| Land Station | | |
| Land Line | | |
| | | |
| Surcharge | | |
| | | |
| TOTAL £ | | |

R/T | * VHF/MF/HF | Duration Mins.
* DELETE WHICHEVER IS NOT APPLICABLE

COASTLINES ABERDEEN

* DELETE IF NOT APPLICABLE

ETA 1700 HOURS STOP REQUEST 500 GALLONS FUEL AND 50 GALLONS LUBOIL ZOMAL =

MASTER +

## Guided writing

### *(B) Writing up the radiotelephone log*

S.S. ..... ESSEX COAST GMML ......
M.V.

DIARY OF THE RADIOTELEGRAPH SERVICE

| DATE AND TIME G.M.T. | STATION FROM | STATION TO | FULL DETAILS OF CALLS, SIGNALS AND DISTRESS WORKING, ETC., AS PRESCRIBED BY RULE 19 OF THE MERCHANT SHIPPING (RADIO) RULES, 1965 | FREQUENCY |
|---|---|---|---|---|
| 12.3.84 | | | QTO Aberdeen bound Leith | |
| 0830 | | | On watch J. Smith. Battery tested – fully | 2182 |
| | | | charged condition | |
| 30/33 | | | Silence period observed | " |
| 35 | | | L.S.W. ON | " |
| 35 | GMML | GND | QTC 1 TR. QSW 2016 K | 2381 |
| 36 | GND | GMML | R 2016 lsn 1856 K // R up | 1792 |
| 37 | GMML | GND | HW? K | 2016 |
| 37 | GND | GMML | QRK 5 QRV K | 1856 |
| 39 | GMML | GND | TR ESSEX COAST/GMML QTO ABERDEEN BND LEITH K | 2016 |
| 39 | GND | GMML | R TR K | 1856 |
| 41 | GMML | GND | Sent nr 1 K | 2016 |
| 41 | GND | GMML | R 1 QTC 1 QRV? K // R QRV K | 1856/2016 |
| 44 | GND | GMML | Recd nr 1 = "QRU QRU? K" | 1856 |
| 45 | GMML | GND | R 1 QRU OUT // R TU OUT | 2016/1856 |
| 46 | | | Resume watch L.S.W. OFF | 2182 |
| | | | | |
| | | | | |

# UNIT XIV SHIPBOARD ELECTRONICS

## Reading comprehension

### *(A) Description of a Marine VHF FM transmitter*

*Page 207*

Microphone – Converts the sound energy into electrical energy.
Amplifier – Amplifies the signal.
Limiter – Limits the amplitude to a predetermined amount.
Pre-emphasis stage – Ensures that the higher audio frequencies are amplified in such a way that these components are not swamped in the noise level inherent in such systems.

| | | |
|---|---|---|
| Oscillator | – | Generates the first carrier frequency. |
| Mixer | – | Mixes the frequency modulated signal with a second carrier. |
| Channel oscillator | – | Changes the frequency to obtain desired transmission frequency. |
| Filter | – | Rejects all unwanted frequencies. |
| Power amplifier | – | Amplifies the signal at the correct frequency. |

## Grammar

### (A) Describing component parts, position and connection

*Exercise* 1.

4. The aerial unit or antenna assembly, which **includes** the reflector, aerial driving motor, synchro-generator, etc. The transmitter and early stages of the receiver are sometimes **mounted** in the aerial unit. With this arrangement the problem arises of keeping the components dry and repairing them, especially in wet weather. The presence of water or moisture in the waveguides, etc., reduces the radiation materially, if not completely and it is therefore common for electrical heating equipment to be **installed**. A thermostat may sometimes be **fitted** to keep the temperature of the apparatus at a constant level. To facilitate the repair or inspection of the aerial, a telephone **connected** to the indicator unit is sometimes **installed** near the aerial. A switch may also be **located** there for switching on and off the driving motor of the aerial. The scanner should, as far as possible, be **situated** well clear of the obstacles almost always present on shipboard, e.g. funnels, derricks, ventilators, etc., to give an uninterrupted sweep of the horizon. It should be **mounted** as high as possible to increase the radar's range; but as the angle of incidence increases, the sea echoes become more pronounced, and this is not desirable.

(NB. This is one possible version, other choices are often possible.)

### (B) Instructions

*Exercise* 1.

2. If the mains lamp does not light, check the mains is connected and the fuses are OK.
3. If the mains lamp lights, select channel and key transmitter.
4. If the transmit lamp does not light, check the power supply to the transmitter.
5. If the transmit lamp lights, select all other channels and key transmitter.
6. If the transmitter does not operate on all channels, check the channel oscillator.

7. If the transmitter operates all channels, check with 2nd transceiver.
8. If the transmitted signals are not received OK, check the Mic/AF section in the transmitter.
9. If the transmitted signals are received, the transmitter is OK.

*Exercise* 2.

(a) Check that the transmit lamp is OK.
(b) Check the aerial.
(c) The associated transmitter.
(d) Check the channel oscillator.
(e) Check the receiver power supplies.

## Reading comprehension

## (B) The echo sounder

*Page 212*

750 metres and 375 metres.

*Page 213*

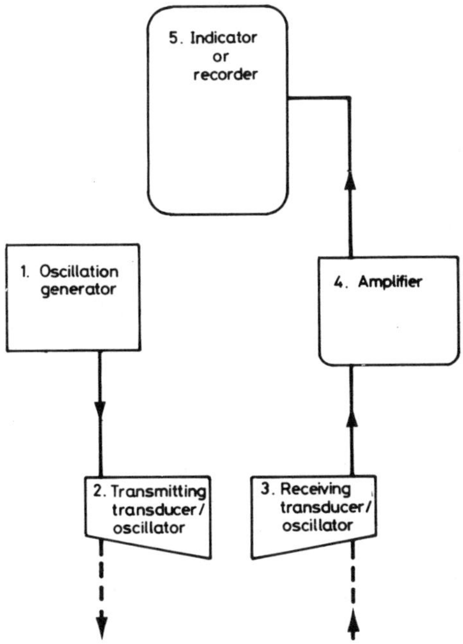


Fig. 14.2 The components of an echo sounding apparatus

*Page 214*

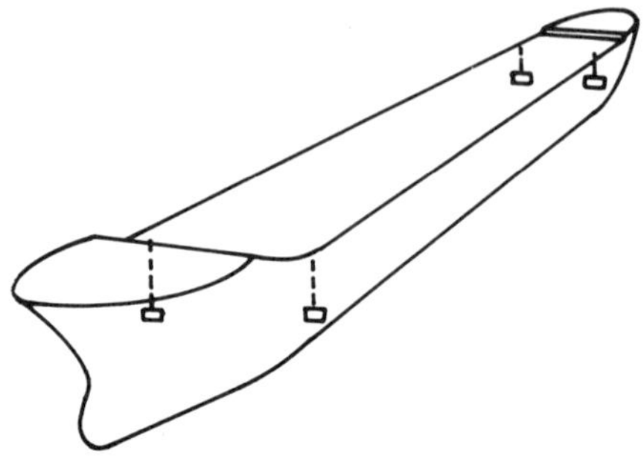

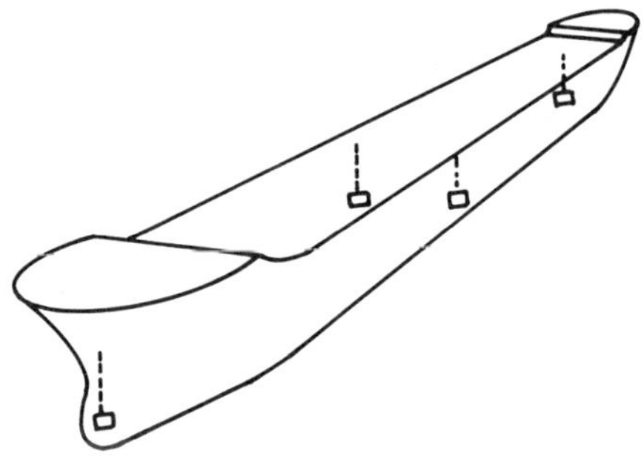

## Applied terminology

### (A) Terms relating to circuit diagrams

1. diode
2. resistor
3. npn transistor
4. capacitor
5. varactor
6. choke/winding of a coil
7. crystal
8. pnp transistor.

### (B) Measurement

*Exercise* 1. *Page 217*
(a) 4Hz. (b) ¼ sec.

*Exercise* 2.

| Formula | Meaning in full |
|---|---|
| $f = \frac{v}{\lambda}$ | frequency equals velocity divided by wavelength. |
| $\lambda = v/f$ | wavelength equals velocity divided by frequency. |
| $v = \lambda f$ | velocity equals wavelength multipled by frequency. |

# APPENDIX II

# TAPESCRIPTS OF LECTURES

## UNIT I

### Lecture: Buoyage

In the IALA Buoyage System 'A', that is to say the system of buoyage adopted by the International Association of Lighthouse Authorities, Cardinal Marks are divided into four groups corresponding to the four points of the compass.

A North Cardinal Mark is black over yellow in colour. It can be either pillar-shaped or spar-shaped. Top marks are two black cones one above the other and with the points upwards. When a light is fitted, it is white and gives a very quick flash or a quick flash.

An East Cardinal Mark is black with a yellow band in colour. It can be either pillar-shaped or spar-shaped. Top marks are two black cones with their bases together. When a light is fitted it is white in colour and gives three very quick flashes every five seconds or three quick flashes every ten seconds.

A South Cardinal Mark is yellow over black in colour. It can be either pillar-shaped or spar-shaped. Top marks are two black cones one above the other and with their points downwards. When a light is fitted, it is white in colour and gives six very quick flashes plus one long flash every ten seconds or six quick flashes plus one long flash every fifteen seconds.

A West Cardinal Mark is yellow with a black band in colour. It can be either pillar-shaped or spar-shaped. Top marks are two black cones one above the other, but with their points together. When a light is fitted, it is white in colour and gives nine very quick flashes every ten seconds or nine quick flashes every fifteen seconds.

Isolated Danger Marks are black in colour with red bands. They can be either pillar-shaped or spar-shaped. Top marks are two black spheres one above the other. When a light is fitted it is white in colour and flashes in groups of two.

Safe Water Marks are red with white vertical stripes. They are spherical in shape, pillar-shaped or spar-shaped. Top marks, which are optional on spherical buoys, are a single red sphere. When a light is fitted, it is white in colour and it either flashes or occults or is isophase.

# UNIT 11

## Lecture: Special duty vessels

There are several types of ship working round ports and channels which are designed to do special jobs to help ships and shipping. Some of these are described briefly in this lecture.

One very useful type of vessel is the tug. Tugs can be divided into four basic types. Some are designed as river tugs for work on rivers. Others are designed as harbour tugs and help ships in and out of ports and harbours. Two other types of tug, which are of growing importance, are coastal tugs and ocean-going tugs. These go out to help ships in difficulty at sea. Tugs must be designed to satisfy three important requirements. They must be stable in all conditions. They must also be manoeuvrable and be powerful enough to move ships of far greater size.

A rather noisy and smaller type of vessel is the dredger. Dredgers are necessary to remove the sand and mud from the beds of channels and harbours. Dredgers are of three main types: they can be either bucket dredgers, which have a series of buckets which go down to the sea bed and scoop up the sand and mud; they can be suction dredgers, which suck up sand and mud like a very large vacuum cleaner; or they can be grab dredgers, which operate like cranes.

Another type of special duty vessel is the icebreaker. Icebreakers are important to shipping because northern ports and channels freeze up in winter. Ships must use these ports all the year round, therefore it is necessary to keep them open. Icebreakers have powerful engines and very strong hulls.

An unusual type of vessel is the lightship. Lightships look like ordinary ships, but they do not have engines, because they are towed into position and then anchored there. They not only have a light, but also a foghorn, a radio beacon and meteorological equipment as well. Most lightships have a crew of approximately twelve.

A very important type of boat is the lifeboat. Lifeboats are of many different types. In the United Kingdom they are manned by volunteers and supported by voluntary donations. Lifeboats must be strong, stable and manoeuvrable and their crew must be well trained.

Finally, there is the pilot launch. Pilot launches are motor boats for transporting pilots to and from ships. They must be seaworthy as pilots go out in all weathers. In the United Kingdom, some port authorities employ their own pilots, but many pilots are employed by Trinity House, which was founded in 1514.

# UNIT III

## Practice 1

During the past twenty years a number of improvements have been made in the design of cargo ships. In the first place, cargo ships are now faster than they were before. Secondly, they are much larger in size. Another improvement has been the improved use of hull space. Finally, there has been a reduction in the time needed to load and unload cargo.

## Practice 2

During the past twenty years a number of improvements have been made in the design of cargo ships. In the first place, cargo ships are now faster than they were before. This is partly because of improved machinery. Secondly, cargo ships are much larger in size. One reason for this increase in size is that it is more economical to run one large ship than two smaller ones. Another improvement has been the improved use of hull space. More room for cargo has been gained by putting the engine room and bridge superstructure at the after end of the ship. Finally, there has been a reduction in the time needed to load and unload cargo. This has been done in a number of ways. One way has been the introduction of better cargo-handling equipment.

## **Lecture:** Improvements in cargo ship design

During the past twenty years or so a number of important improvements have been made in the design of cargo ships.

In the first place, cargo ships of all types are now faster than they were before. This is partly because of improved machinery and partly because of improvements in hull design.

Secondly, cargo ships are much larger in size nowadays. One reason for this increase in size is that it is more economical to run one large ship than two smaller ones. Another reason is that ports all over the world are constantly being improved. A third reason is that improvements in ship construction make it possible to build larger ships.

Another improvement in ship design has been the improved use of hull space. More room for cargo has been gained by putting the engine room and bridge superstructure at the after end of the ship. More room for passengers in passenger ships has been gained by reducing the number of classes.

Finally, there has been a reduction in the time needed to load and unload cargo. This has been done in a number of ways. One way has been the introduction of better cargo-handling equipment. Another way has been the introduction of specialized vessels, which are designed for loading and unloading a particular type of cargo. For example, bulk carriers and container ships.

# UNIT IV

## **Lecture:** General purpose manning

### *Introduction*

In this short lecture I want to talk about a system of crew organization known as General Purpose manning. GP manning, as it is often called, is a system used particularly on tankers and container ships. First, I want to talk about some of the changes from the traditional system and then I want to look at some of the reasons for the development of this type of system. Lastly, I want to give you some of the advantages.

### *Part 1*

In front of you is a diagram showing the possible organization of a tanker with a general purpose crew. As you can see, there are changes in the organization of Petty Officers and ratings and some well-known names, such as Bosun, Carpenter and Donkeyman, have disappeared. In this system general purpose ratings take the place of greasers and seamen in traditionally manned ships. Engine room and deck ratings join together to form a work force capable of working on deck and in the engine room. In ships with this type of crew, GP petty officers are capable of doing the jobs of Bosun, Carpenter and Donkeyman, and in tankers, doing the job of Pumpman as well. The work of the GP crew is organized by a planning committee. The planning committee is usually made up of the Master, who acts as Chairman, the Chief Engineer, the Chief Officer, the Second Engineer, the Catering Officer and the Chief Petty Officer. The Radio Officer acts as Secretary. The committee usually meet once a week to plan the programme of work. Daily meetings can also be held to make last-minute changes, when necessary.

### *Part 2*

There have been several reasons for the change to the GP system of manning by some companies. In the first place, technical developments have led to an increase in the amount of automation on board ship nowadays. With more automation there is, therefore, a need for less manpower; that is, fewer people are necessary. But the ratings need to be better trained in order to operate more complicated machinery. In the second place, there have been changes in the way merchant ships are used. Because of economic reasons, it is now necessary for ships to spend more time at sea and less time in port. It is therefore necessary to be able to change the crew in order to allow leave and some ships now have more than one crew assigned to them, particularly in the case of tankers. The GP system makes the changing of a crew easier and more economic.

### *Part 3*

The GP system of manning has a number of advantages. These can be divided into advantages for the company and advantages for the GP rating. Firstly, the advantages for the company. This system reduces the size of the crew by combining the deck and engine room ratings into one work force. Another advantage is that the work force can be used more efficiently. For example, when the weather is fine, it can work on deck, when the weather is bad, it can work on maintenance in the engine room. Secondly, the advantages for the GP rating. Under this system he is better paid because he is more qualified. In addition there is also the opportunity of earning extra money for overtime. Another advantage is that a smaller but better qualified crew can expect superior accommodation. Generally speaking the GP system of manning leads to greater job satisfaction.

# UNIT V

## **Lecture:** Different types of rope

### *Introduction*

A large number of different types of rope are used on board ship, and it is important for every sailor to know their characteristics so that the right rope can be used for the right job. Ropes can be divided into three basic types: natural fibre rope, which is made from the fibres of different plants; synthetic fibre rope, which is made from materials such as nylon; and wire rope, which is made from strands of steel wire.

First let us look at the different types of natural fibre rope. A well-known rope of this type is Manila. Manila rope is made from the fibres of a plant which grows in the Philippine Islands of the Pacific. It is strong and flexible, but rather expensive. It is used for a number of jobs connected with cargo-handling and mooring. Because manila rope is expensive, sisal rope is often used in its place. Sisal comes from a plant which grows in the USA and Russia. It is less strong and less flexible than manila rope, but it is cheaper. It is used for moorings and lashings. Another type of rope is hemp rope. Hemp comes from a plant which grows in the USSR, Europe and North America as well as in China and India. It is strong and flexible and does not shrink or swell after contact with water. Because of this it is used on sailing boats. Coir ropes are made from coconut fibres. They are very buoyant and very elastic, but they rot easily when they are wet. They are sometimes used for mooring and towing lines. The cotton plant grows in the southern part of North America. Cotton rope is both strong and flexible, but it is very expensive and therefore not used on merchant ships. Because it looks nice, it is often used on yachts and pleasure boats.

Natural fibre ropes have now largely been replaced by synthetic fibre ropes. Synthetic ropes have many advantages. They are strong and elastic and they are resistant to the action of water. Nylon rope is the strongest and the most elastic of all the synthetic fibre ropes. It is used for mooring and handling cargo. Terylene rope has the highest melting point. It melts at a temperature of 260°C. It is also strong and elastic. It is mainly used on yachts. Another type of synthetic fibre rope is polypropylene rope. It has the lowest melting point of all synthetic ropes and is used for log lines and halyards.

Wire rope is made of steel. It is usually galvanized to stop it from rusting. It is very strong and elastic, but not as flexible as other types of rope. Large wire ropes are very heavy. Wire rope has many uses on board ship, particularly for standing rigging, mooring lines and for cargo-handling.

# UNIT VI

## **Lecture:** Advantages and disadvantages of containerization

### *Introduction*

In this lecture I want to discuss some of the advantages and disadvantages of containerization as a method of transporting general cargo. First, I want to look at some of the advantages. These refer to both cost and time. Secondly, I want to look at some of the disadvantages, which also refer to cost, and at, what might be called, inconvenience.

(*Note:* The text of Part 1 of this lecture appears in the Note-taking Section of Unit 6.)

### *Part 2*

Now some of the advantages with reference to time. Time is an important factor and is directly related to cost, because in any service time costs money. Containerization reduces handling and less handling means less time is necessary for transporting the goods. Time is further reduced by the easier stowage of containers compared with non-containerized general cargo. Containers are of a standard shape and dimensions, therefore they are easily stowed. In addition to easier stowage, containers are more easily loaded and unloaded. This is done quickly by special cranes. Time is also reduced by the fact that less handling means less paperwork. Because the goods remain in the container they do not have to be checked when loaded and unloaded into and off the ship.

### Part 3

But containerization also has a number of disadvantages. These concern cost and, what might be called, inconvenience. With reference to cost, a containerized transport system is expensive to establish. The ship owner has not only to buy specially designed container ships, but also the containers to go with them. However, these can be hired and some shippers provide their own containers. Special port facilities, such as special cranes, must also be provided by the port authority. These are expensive to build and beyond the means of some countries and port authorities.

### Part 4

With reference to inconvenience, containerization causes a number of problems. Not all cargo can be put into containers. However, the percentage of non-containerizable cargo falls each year as new types of container are introduced. Another inconvenience is that different types of container are needed for different journeys and different times of year. It may also be difficult for a small shipping company to fill a container to capacity with compatible cargo. It is uneconomical to carry half full containers in a ship. However, these difficulties have largely been overcome by the co-operation of all those involved. Finally, a containerized system of transport depends on good road and rail links with the port and the surrounding countryside. In some countries these have not yet been developed.

# UNIT VII

## Shipping forecast at 0640 hours

Gale warnings are in operation in sea areas Forties, Dogger, Fisher and German Bight.

The general synopsis at midnight last night. A depression of 988 millibars which was positioned at 60° North 10° East is expected to move north-east. A depression of 1004 millibars which was positioned at 56° North 19° West is expected to move east and to be centred over the south of Scotland by midnight tonight.

The sea area forecast is as follows:

*Viking:* Wind north-westerly force 6, becoming north-westerly force 5 to 6. Periods of rain, moderate to good visibility.

*Forties:* Wind north-westerly force 6 to 8, moderating to north-west force 3 to 5, rain showers, moderate to good visibility.

*Cromarty:* Wind north-westerly force 3 to 5, veering to south-east force 6, rain, good to moderate visibility.

*Dogger, Fisher, German Bight:* Wind north-westerly force 6 to 8, moderating to force 4 to 5, periods of rain, good visibility.

*Humber, Thames, Dover:* Wind westerly force 3 to 5, force 6 in Dover, backing south-westerly force 5 to 7 in Thames, rain, good visibility.

*Wight, Portland, Plymouth:* Wind westerly force 4 to 6, rain turning to drizzle with mist, poor visibility.

*Biscay:* Wind westerly force 3, cloudy, good visibility.

*Finisterre:* Wind north-easterly force 3 to 5, cloudy, good visibility.

*Sole, Lundy, Fastnet, Irish Sea, Shannon:* Wind south-westerly force 4 to 6, rain at first, drizzle with mist later, good to poor visibility.

*Rockall:* In the south of the region the wind will be south-westerly force 4 to 6 and in the north south-easterly force 3 to 5.

*Malin:* Wind variable force 3 with winds becoming force 5 to 7 at the centre of the low.

*Hebrides, Minches:* Wind variable, becoming easterly force 3 to 6.

*Fair Isle:* Wind northerly force 6, becoming easterly force 6, cloudy, good visibility.

*Bailey:* Wind south-easterly force 5, cloudy, good visibility.

*Faeroes:* Wind northerly force 3 to 5, becoming easterly force 3 to 5, cloudy, good visibility.

*SE Iceland:* Wind variable force 3, becoming south-easterly force 4 to 6, cloudy, good visibility.

## *Coastal reports for 0400 hours*

*Wick:* West force 2, 16 miles, 1016 millibars, rising.

*Bell Rock:* West force 5, cloudy, 22 miles, 1016 millibars, rising slowly.

*Dowsing:* Wind west-north-west force 4, cloudy, 5 miles 1019 millibars.

*Galloper:* Wind west force 4, 5 miles, 1021 millibars, steady.

*Royal Sovereign:* Wind westerly force 4, 16 miles, 1022 millibars, steady.

*Portland Bill:* Wind west by south force 3, 22 miles, 1022 millibars, falling.

*Scilly:* Wind south-west force 4, light rain during the past hour, 16 miles, 1023 millibars, falling.

*Valentia:* Wind south-west force 4, light rain, 6 miles, 1017 millibars, falling.

*Ronaldsway:* Calm, light rain during the past hour, 9 miles, 1017 millibars, falling.
*Prestwick:* Wind south-easterly force 2, light rain, 13 miles, 1017 millibars, falling.
*Tiree:* Wind east-south-east force 4, 27 miles, 1015 millibars, falling.

# UNIT VIII

## Lecture: Boilers

Boilers are used on board ship for producing steam. This steam may be used for driving the main engines, when steam turbines are fitted, or for driving auxiliary machinery such as the windlass. There are two basic types of boiler in use in ships: the Scotch boiler, which is a type of smoke-tube boiler, and the water-tube boiler. Both are illustrated in the diagrams in front of you.

Now if you look at the diagram of the Scotch boiler you will see that it consists of a cylindrical steel shell, which contains a furnace at the bottom. Two or more furnaces may be fitted, depending on the size of the boiler. The furnace is connected to a combustion chamber, situated in the middle part of the boiler. Leading from this combustion chamber to a smoke box are a large number of horizontal tubes. Some of these are used for supporting the middle section of the boiler and the combustion chamber. Connected to the smoke box are the uptakes for the waste gases. The furnace, the combustion chamber and the tubes are all surrounded by water.

The gases are heated in the furnace by the heat from flames and from the sides of the furnaces. The hot gases then enter the combustion chamber. Like the furnace, this is also surrounded by water and so again provides for the generation of steam. From the combustion chamber the gases pass through the smoke tubes, where the surrounding water is exposed to a greater area of heated surface. After leaving the smoke tubes the gases enter the smoke box and after that the uptakes. The steam generated collects in the top of the boiler.

The Scotch boiler is usually of both riveted and welded construction. It is strong and is capable of operating with poor quality feed water. But there are more efficient types of boiler which can generate more power, therefore Scotch boilers are now mainly used for auxiliary purposes on board ship.

Water-tube boilers have replaced Scotch boilers for generating steam for main engines. From your diagram you will see that they have a steam drum at the top, which is partly filled with water, and water drums at a lower level. These drums are connected by banks of tubes, which also contain water. The furnace is located at the bottom and the whole system is contained in a fire-proof casing. Downcomer tubes are placed outside the gas system to act as feeders to the water drums.

Gases are heated in the furnace and pass upwards through the banks of tubes, transferring their heat to the water in the tubes. Because the steam drum provides a reservoir of relatively cool water, convection currents are set up causing the water to circulate round the system. The banks of tubes offer a large surface area to the radiant heat of the furnace gases. This makes them very efficient. Superheaters are added to the system to increase its efficiency. These are located between the rows of tubes.

Various valves and gauges are fitted to the boilers. Those attached directly to the pressure parts of the boiler are known as boiler mountings. For a water-tube boiler these include the following: safety valves, which are needed to release any excess steam from the boiler; a main stop valve in order to control the passage of steam to the engines; feed check valves to control the final entry of water into the boiler; water level indicators to show the level of water in the boiler and thermometers for showing the temperature inside the boiler. In order to be able to drain water from the system drain valves are fitted. Salinometer valves are also fitted to allow samples of water to be drawn off for testing. Chemical dosing valves are also necessary so that chemicals can be added directly into the boiler.

# UNIT IX

## Practice 1

(a) three point one four two
(b) nine minus three equals six
(c) six kilometres per hour
(d) five per cent
(e) the square root of *x*
(f) one foot equals zero point three zero four eight metres
(g) ninety degrees
(h) *x* is not equal to *y*
(i) fifteen plus thirty equals forty-five
(j) *a* divided by *b* equals *c*
(k) in the ratio of four to three
(l) *x* to the power four
(m) less than two per cent
(n) sixty degrees Fahrenheit is approximately equal to fifteen degrees Centigrade
(o) thirty-two degrees ten minutes
(p) force is proportional to mass multiplied by acceleration
(q) plus or minus zero point one five millimetres
(r) *x* is greater than *y*
(s) twelve thousand and twenty-five
(t) twelve point two five

(u) one radian is approximately equal to fifty-seven point three degrees
(v) $x$ to the power minus $n$
(w) one hundred and two kilonewtons per square metre
(x) the cube root of $n$ minus one

# UNIT X

## **Lecture:** Maintenance schedule of marine diesel engine

Engine builders supply detailed instructions on the operation and maintenance of their machinery so that regular maintenance work can be carried out and breakdowns can be kept to a minimum. These instruction manuals are usually kept by the Chief Engineer, but they are made available to all members of the engine-room staff. The intervals at which an engine and its parts must be inspected will vary from make to make and will depend on the use the engine has been put to, and therefore the brief outline which follows is meant only as a general guide.

At frequent intervals, fuel pumps should be examined and adjusted if necessary. When the engine is running, this will be shown by comparing engine indicator cards and by exhaust temperatures. Pistons should also be examined frequently for cracks.

At intervals of six weeks, the fuel valves should be taken out and carefully inspected. Atomizers and filters can be washed with clean paraffin and then dried in a warm place. Cleaning rags must not be used because they leave behind small pieces of fluff, which may block up holes. Valve seats should also be tested and if they are pitted or scratched, the surface should be reground.

If possible, the upper piston rings should be examined at intervals of one month during the first six months' service. After that inspection periods can be extended so long as their condition continues to be satisfactory.

At intervals of six months the upper pistons, if cooled, must be inspected for deposits of carbon in cooling spaces and cooling pipes. When new piston rings are fitted, care must be taken to ensure there is sufficient clearance to allow for the expansion of the rings. Exhaust belts and manifold must also be examined and excessive carbon deposits removed. All carbon deposits should also be removed from cylinder ports. Cylinder liners must be examined externally for deposits of scale. If these deposits cannot be removed by flushing with water, then the liner must be removed for cleaning. The liner should also be measured for wear and renewed, if the limit for wear has been reached. The clearances of connecting-rod top and bottom ends should also be examined every six months and adjusted if necessary. In addition, lubricating-oil sumps and tanks should be cleaned of sediment.

At intervals of one year the manoeuvring gear must be examined for wear at the joints of levers and rods. The alignment of the crankshaft should be checked and any incorrect alignment corrected. The main bearings must be examined and readings taken for wear. The clearances of all crankshaft bearings must be maintained at the figure recommended by the makers. Finally, starting air piping and air bottles must be cleaned and steamed out, and the lubricating oil system thoroughly examined and cleared of deposits.

It must be emphasized that the above-mentioned parts are only some of the items which must be regularly maintained to ensure the efficient working of the machinery.

# UNIT XI

## Practice 1

A life-boat should contain the following equipment:

1 Rudder and tiller
1 Mast
2 Sails
5 Oars
Double lifeline round boat
1 Exposure cover
2 Buckets
6 Crutches
1 Sea anchor
1 Water-tight box
2 Boxes of matches
1 Bailer
2 Hatchets
1 Compass
1 Gallon of storm oil
2 Smoke signals
1 Lamp
2 Plugs
1 first-aid kit
2 Boat hooks
1 Locker
2 Painters
1 Signalling mirror
1 Waterproof torch
1 Spare set of batteries
2 Buoyant heaving lines
1 Fishing line with 6 hooks
6 Hand flares
Food and water
1 Jack-knife
1 Whistle
1 Spare torch bulb
4 Parachute signals
1 Manual pump
1 Copy of the Deapartment of Trade Rescue Table

# UNIT XII

## Exercise 1 The phonetic alphabet

*Exercise* 1.

Alpha

Bravo

Charlie
Delta
Echo
Foxtrot
Golf
Hotel
India
Juliet
Kilo
Lima
Mike
November
Oscar
Papa
Quebec
Romeo
Sierra
Tango
Uniform
Victor
Whisky
X-ray
Yankee
Zulu

## Exercise 3 The figure code

*Exercise* 3.

Nadazero
Unaone
Bissotwo
Terrathree
Kartefour
Pantafive
Soxisix
Setteseven
Ocktoeight
Novenine
Stop
Decimal

## Practice 1

| | |
|---|---|
| AE | Alpha Echo |
| BF | Bravo Foxtrot |
| CB | Charlie Bravo |
| CB6 | Charlie Bravo Soxisix |
| DX | Delta X-ray |
| HW | Hotel Whisky |
| CP | Charlie Papa |
| ED | Echo Delta |
| EL | Echo Lima |

## Practice 2

'Mayday, Mayday, Mayday.
This is Maritime, Maritime, Maritime.
Mayday Maritime,
Interco Alfa Nadazero Unaone Pantafive Ushant Romeo Kartifour Nadazero Delta X-ray, Over.'

'Mayday, Mayday, Mayday
This is Golf Yankee Sierra Hotel Golf Yankee Sierra Hotel
Golf Yankee Sierra Hotel

Mayday, Golf Yankee Sierra Hotel
Interco Lima Pantafive Kartifour Bissotwo Pantafive November
Golf Nadazero Unaone Soxisix Terrathree Terrathree Whisky
Charlie Bravo Soxisix, Over.'

## **Lecture:** Authorities giving aid to ships and aircraft

In this lecture I want to talk briefly about the authorities which will give assistance when a ship or aircraft is in distress off the coasts of the United Kingdom.

Firstly, there are the Coast Radio Stations. There are eleven of these and they are operated by the Post Office. They all keep a continuous watch on the distress frequencies of 500 kHz and 2182 kHz except Oban, which keeps watch on 2182 kHz only. When a radio distress signal is received by a Coast Radio Station, it is transmitted to ships at sea. In addition, different authorities ashore are also notified.

Secondly, there is Her Majesty's Coastguard. This authority is responsible for initiating, that is starting, and also co-ordinating the search and rescue measures for all vessels in distress off the coast of the United Kingdom. The area over which this responsibility extends is bounded by latitude 43° and 68° North, by longitude 30° West and by the coastline of western Europe or 8° East, whichever is nearer to the coastline of the United Kingdom. This area can be reached by long-range aircraft capable of operating up to 1000 miles from the shore. There are a large number of these stations around the coast, some keep a continuous radio and visual watch throughout twenty-four hours, others keep a visual watch during busy shipping periods and bad weather.

Thirdly, there is the Royal National Lifeboat Institution. This is a private organization, which is supported entirely by voluntary contributions of money. It maintains 137 lifeboats. It also has a fleet of about 108 inshore rescue boats. Each lifeboat has a radio set which operates on the distress frequency of 2182 kHz. Lifeboats also have other radio equipment so that they can communicate with other vessels, shore stations, aircraft and helicopters.

In the fourth place, there is the Royal Navy. The Royal Navy gives assistance by means of ships, aircraft and helicopters.

In the fifth place, there is the Royal Air Force. Although responsible for assisting military and civil aircraft, the Royal Air Force will help ships in distress by means of aircraft when this is possible.

In the sixth place, there are Air Traffic Control Centres. These are often the first to receive information about aircraft in distress. All commercial and many private planes are able to communicate with these centres by radio. Planes may be asked to keep a look-out along or near their normal routes, and to report the position of any casualties they see.

In the seventh place, there are the ocean-going tugs. These use their own radios to learn of ships in distress. They then go out to the ship offering assistance in the hope of gaining salvage money.

Lastly, there are the Officers of the Fishery Departments who inform the Coastguard of fishing vessels which are missing or which are overdue.

# UNIT XIII

## Radiotelephony

'Stonehavenradio Stonehavenradio Stonehavenradio
This is 'Essex Coast' 'Essexcoast'
Callsign Golf Mike Mike Lima
I have one radiotelegram and Tango Romeo for you
Listen 2016 – over.'

'Essexcoast Essexcoast
This is Stonehavenradio
Romeo 2016 Listen 1856 – Over'
(Meaning – I agree to 2016KHz.
Listen for me on 1856KHz).

'Romeo up'
(Meaning – received and understood,
changing frequency now).

Ship now changes transmitter frequency to 2016KHz, and receiver to 1856KHz.
Coast station changes transmitter to 1856KHz and receiver to 2016KHz.

'Stonehavenradio Stonehavenradio
This is 'Essexcoast' 'Essexcoast'
How are you receiving me – Over'.

'Essexcoast
This is Stonehavenradio
Receiving you well – Go ahead – Over'.

'Stonehavenradio
This is Essexcoast
Tango Romeo begins 'Essexcoast'
I spell Echo Sierra Sierra Echo Xray
Charlie Oscar Alpa Sierra Tango
Golf Mike Mike Lima leaving
Aberdeen Bound Leith – Over'.

'Essexcoast
This is Stonehavenradio
Received your Tango Romeo –
Go ahead – Over'.

'Stonehavenradio
This is Essex Coast
Message begins – prefix Mike Sierra
Golf from Essexcoast number one
number of words sixteen date
twelfth time 0830 Account code
number Golf Bravo Zero eight
Address begins – Coastlines Aberdeen
Text begins – ETA I spell Echo Tango
Alpha in figures 1700 hours Stop
Request in figures 500 gallons fuel
and in figures 50 gallons luboil code
group Zulu Oscar Mike Alpha Lima
Signature – Master.
Telegram ends I collate
(Repeat message through at normal
speech rate)
Telegram ends – Over.'

'Essexcoast
This is Stonehavenradio
Received your number one
I have nothing for you – Out.'

'Stonehavenradio
This is Essexcoast
I have no further traffic – Out.'

# UNIT XIV

## **Lecture:** A VHF transceiver

In front of you is a block diagram of an Argonaut S VHF transceiver, which is capable of communicating up to a distance of about 50 miles. The block is split into two basic units: the transmitter to the left hand side of the diagram, and the receiver to the right. There is a common part to both the transmitter and receiver known as the frequency synthesizer.

Tracing the transmitter side first, under the block marked control unit there is a microphone. Audio information produced by the voice is picked up by the microphone and changed into an electrical signal. The signal is then amplified, before being pre-emphasized. This stage amplifies the higher frequency components in the voice more than the lower frequency components. There is then a filter based on 300 – 3000 Hz., the main components in voice frequency, and a limiter to ensure the amplitude is not too great. As this is a frequency modulated system, the next stage contains an 11.7 MHz. oscillator, which is frequency modulated with the audio information. This signal is then applied to a mixer, marked unit 1300. The other input to this mixer being the channel and band information, this information coming from the frequency synthesizer. The frequency synthesizer produces the correct frequency for the mode of emission, but note, this is going to produce half the final output radiated frequency. The signal from the output of the mixer is then filtered to remove unwanted components. It is amplified and then goes to another stage within unit 1500, which again amplifies the signal and doubles the frequency. This frequency is then the final radiated frequency. The signal is then amplified further by a driver stage and the PA, that is, the power amplifier. The final transmitted output power is thus raised to approximately 20 Watts. The signal is filtered and then applied to the transmitter aerial. In addition to this, there is a keying device, whereby you must enable the transmitter to transmit as it is not running all the time, and is usually actuated by a button on the handset of the transceiver unit itself. When this button is not operated the receiver circuit is open.

Note that there is a common aerial system. From the aerial filter unit just beside the antenna, or aerial system, there are two paths for the received signal, one marked simplex, the other marked duplex. Duplex is not strictly necessary for basic communications between ships, most communication being done in a simplex mode. This simply means that the transmit and receive frequencies are the same. For commercial working through a coast station for radiotelephony calls, it is more normal to select a duplex channel, where the transmit and receive frequencies are not the same. On selecting duplex from the control unit, it is not necessary to key the transmitter. So, following either path, depending on whether simplex or duplex is selected, the signal is amplified, using an RF, or radio frequency, amplifier. The signal is applied to a mixer again which mixes the radio frequency input with a signal from the frequency synthesizer. The two signals beat together and the difference frequency, 10.7 MHz., is put through an IF amplifier, or intermediate frequency. A second mixer takes a second local oscillator signal and the 10.7 MHz. signal and produces a second IF at 455 KHz. The signal is again limited before being applied to the discriminator, which removes the audio frequency information from the carrier signal. The signal is de-emphasized through this de-emphasis stage, which brings the high frequency components back to their original level. This audio information is then amplified using the AF amplifier and applied to the loudspeaker, shown on the control unit, unit 1800. In addition, there is a system known as squelch, which means that the receiver audio frequency amplifier is not operational until a carrier signal is present. This ensures that unwanted noise, present on VHF, is not obtrusive.

The unit on the bottom of the diagram is the power supply unit. The mains input at, in this case 220V 50Hz., is seen on the left hand side marked off/on, which is the mains switch. This should also be protected by fuses. The power supply takes this 220V ac, rectifies it to dc, and distributes certain supplies to each unit in the transceiver. In this case, we have got +28V, +20V, −18V and −24V. These are the supplies which the transceiver uses. Not all the lines are drawn in in order to keep the diagram simple. The regulator indicated is the voltage regulator by which the supplies are stabilized. The control unit, unit 1800, as well as having the microphone and loudspeaker, also has a low/high switch. This controls the amount of power needed for transmission: minimum power on the low position being 1 Watt, and the maximum position being 20 – 25 Watts, depending on the unit. In addition to the loudspeaker, there is a squelch off/on switch which controls this circuit. There is also an off/on switch for the control unit in addition to the one on the power supply. The dual watch enables the receiver to switch between selected frequencies, normally channel 16 and a channel selected using the channel select switches. These switches are marked 30.